NORTH AMERICA
IN THE ANTHROPOCENE

North America

in the Anthropocene

ROBERT WILLIAM SANDFORD

RMB

RMB | Rocky Mountain Books Ltd.
rmbooks.com
@rmbooks
facebook.com/rmbooks

Cataloguing data available from Library and Archives Canada
ISBN 978-1-77160-180-1 (hardcover)
ISBN 978-1-77160-181-8 (electronic)

Printed and bound in Canada by Friesens

Distributed in Canada by Heritage Group Distribution and in the U.S. by Publishers Group West

For information on purchasing bulk quantities of this book, or to obtain media excerpts or invite the author to speak at an event, please visit rmbooks.com and select the "Contact Us" tab.

RMB | Rocky Mountain Books is dedicated to the environment and committed to reducing the destruction of old-growth forests. Our books are produced with respect for the future and consideration for the past.

We acknowledge the financial support of the Government of Canada through the Canada Book Fund and the Canada Council for the Arts, and of the province of British Columbia through the British Columbia Arts Council and the Book Publishing Tax Credit.

For Gudmund Hernes,
who is an inspiration to all who care about water
and are concerned about climate.

CONTENTS

CHAPTER 1

Transforming Our World:
Water and Sustainability for All

The timely availability of fresh water has for decades been recognized as a global concern. There is not enough water to support our constantly growing populations and sustain all the uses to which we want to put this precious resource. In addition to matters of availability and quality, we now recognize that the world will soon be redefined by changing precipitation patterns associated with an increase in the mean temperature of our planet's atmosphere. There are going to be winners and losers – places that will remain habitable and places that will not. The geography of human presence on the planet is about to change, and such change is unlikely to occur without conflict. While there are precedents to suggest that outright warfare specifically over water can be avoided, solving the problem of inequitable water supply and reducing the tensions that persistent water shortages create will not be easy nor will it be cheap.

Among the many reports published by the United Nations in the lead-up to the Paris climate conference in 2015 was one published by the United Nations University Institute for Water, Environment and Health, which warned that without large, new, water-related investments, many societies worldwide will soon confront rising desperation and conflict over life's most essential resource. Presenting their report, entitled *Water in the World We Want*, at UN headquarters in New York City, officials of UN University and the UN Office for Sustainable Development said unmet water goals threaten many regions of the world and form a barrier to key, universally shared ambitions, including stable political systems, greater wealth and better health for all.

Water in the World We Want argued that continued stalling, coupled with population growth, economic instability, disrupted climate patterns and other variables, could reverse hard-earned development gains and preclude meaningful development that can be sustained into the future. The report underlined that all current water management challenges will be compounded one way or another by hydro-climatic change and increasingly unpredictable weather. The report noted that historical predictability – what is called relative hydrological "stationarity" – provided the necessary certainty, not only to design structures to withstand winds of a certain speed, snow of a certain weight and rainfalls of certain intensity and duration but also when to plant crops and how big to build storm sewers. Because of warming generated by changes in the composition of the Earth's atmosphere, however, the relative stability of the global hydrological cycle has been lost. The consequence is that the management of water in all its forms in the future will involve a great deal more uncertainty than it has in the past.

In a more or less stable hydro-climatic regime, the report observed, you are playing poker with a deck that you know and, thus, can bet on risk accordingly. The loss of stationarity is playing poker with a deck in which new cards you have never seen before keep appearing more and more often, ultimately disrupting your hand to such an extent that the game no longer has coherence or meaning. People, unfortunately, do not have the luxury of living without water, and when faced with a life or death decision, they tend to do whatever they must to survive. "In this manner," the report noted, "changes in funda-mental hydrology are likely to cause new kinds of conflict, and it can be expected that both water scar-city and flooding will become major transboundary water issues."

Water in the World We Want further noted that within ten years, researchers predict 48 coun-tries – 25 per cent of all nations on Earth, with an expected combined population of 2.9 billion – will be classified as either "water-scarce" (1000 to 1700 cubic metres of water per capita per year) or "water-stressed" (1000 cubic metres or less). By 2030, overall global demand for freshwater could exceed supply by 40 per cent, with the most acute problems oc-curring in warmer, low-resource nations with young, fast-growing populations.

An estimated 25 per cent of the world's major river basins run dry for part of each year, the report noted, and new conflicts are likely to emerge as more of the world's rivers become further heavily abstracted so that they no longer make it to the sea. Meanwhile, the magnitude of floods in Pakistan and Australia in 2010, and on the Great Plains of North America in 2011 and 2014, "suggests that the destruction of upstream flood protection and the failure to provide adequate downstream flood warning will enter into global conflict formulae in the future." The report cited the rising cost of world flood-related damages: US$53-billion in 2013 and more than US$312-billion since 2004.

Published in the run-up to the UN member states' adoption of universal post-2015 sustainable development goals, *Water in the World We Want* provided an in-depth analysis of ten countries to show how ensuring reliable water supply and universal sanitation services offers a rapid, cost-effective way to achieve sustainable development. The countries included in the study covered the full range of economic development: Bangladesh, Bolivia, Canada, Indonesia, Republic of Korea, Pakistan, Singapore, Uganda, Vietnam and Zambia. Based on the national case studies, the report prescribes country-level steps for achieving the global water targets.

The report cautioned, however, that the success of global efforts to achieve sustainable development goals for water at the scale required will largely entail cracking down on widespread corruption in the water sector, particularly but not exclusively in developing countries. The report noted that, in many places in the world, corruption is resulting in the hemorrhaging of precious financial resources, siphoning off an estimated 30 per cent of funds earmarked for water and sanitation-related improvements.

The report underscored the need for clearly defined anti-corruption protocols enforced with harsh penalties. Given accelerating Earth system changes and the growing threat of hydro-climatic disruption, corruption that adds to the cost of water-related improvements threatens the stability and very existence of some nation-states, ultimately affecting everyone. The report went so far as to claim that corruption at any level is not just a criminal act locally in its own right; in the context of sustainable development, corruption could be viewed as a crime against all of humanity.

The *Water in the World We Want* report also noted that the world's water and waste water infrastructure maintenance and replacement deficit is accumulating at a rate of $200-million per year, with $1-trillion now required in the USA alone. To finance its recommendations, the report says that in addition to plugging the leakage of funds to corruption, $1.9-trillion in subsidies to petroleum, coal and gas industries should be redirected by degrees. The estimated global cost to achieve post-2015 sustainable development goals in water and sanitation development, maintenance and replacement is

US$1.25-trillion to $2.25-trillion per year for 20 years, a doubling or tripling of current spending translating into 1.8 to 2.5 per cent of global GDP. The resulting benefits would be commensurately large, however: a minimum of $3.11-trillion per year, not counting health care savings and valuable ecosystem service enhancements.

The report recommended that national governments make sustainable advancements in water, waste water and sanitation management, supported by a dedicated and independent, arm's-length water agency, a high-level policy priority. The document also observed that decisions for managing water at all scales must be informed by evidence, accounting for the multiple roles, uses and demands on water and disposal of human waste and waste water, as well as the way in which the distribution of water resources is changing and is expected to continue to change over time and space.

Capacity development must be nested within, and form a pillar of, institutional reform at all scales within a country, with an emphasis on transferable skills that can be used for sustainable development across all areas and goals. The report also recommended that governments, supported by relevant stakeholders, must commit to timely and transparent monitoring and reporting on sustainable development indicators to monitor progress and hold the global community mutually accountable. There must also be a national commitment to universal access to water and sanitation, linked to waste treatment and management and delivered through nationally coordinated and monitored multi-stakeholder response systems. The report also noted that the world can no longer ignore nature's own needs for water to maintain planetary biodiversity-based Earth system function, and thus the document enjoined national governments to understand nature's need for water and meet that need through broader water management objectives and appropriate action.

The report recommended that national water governance and management include a requirement to balance supply and demand at the sub-basin level for sustainability and disaster risk reduction, while recognizing and protecting downstream users. Common disaster risk reduction targets, the report noted, need to be formally incorporated into post-2015 water- and sanitation-related sustainable development goals. These targets must permit the tailoring of actions to national realities.

The report further argued that the agriculture sector must be held more accountable for water-use efficiencies and other system refinements that limit water demand, while maintaining or increasing productivity, ensuring that women and small-scale farmers are provided with the knowledge and technology to be able to play their part, thereby increasing income above poverty thresholds. The energy sector, too, must be held accountable for water efficiencies and a transition to clean energy, including hydropower and geothermal, which does not compromise water quality, environmental integrity, community access or disaster mitigation.

The *Water in the World We Want* report was also very clear about the importance of involving the private sector in the global effort to achieve sustainable development goals with respect to water. Water-dependent companies have a key role to play, the report noted, in financing and implementing sound water, sanitation and waste water management strategies and must step up to the plate or risk significant losses. This is no longer simply corporate social responsibility; it has become sound economic strategy. National governments, transnational corporations and international institutions must work together to identify and implement strategies to equitably develop existing resources and direct them toward greater efficiency through integration of inter- and intra-sectoral activities that take advantage of economies of scope and scale. Subject to rigorous due diligence, national governments must identify, explore and utilize new mechanisms for financing, including various tax and royalty regimes, as well as private and public sector partnerships.

Dr. Zafar Adeel, director of the United Nations University Institute of Water, Environment and Health, and Jong Soo Yoon, head of the UN Office for Sustainable Development, stated that the report filled a critical gap in understanding the complexities associated with water resources and their management but also provided substantive options that will enable the world to move forward with the global dialogue on humanity's relationship to water and sustainability.

THE UN'S 2030 *TRANSFORMING OUR WORLD* SUSTAINABLE DEVELOPMENT AGENDA

As reports such as *Water in the World We Want* make clear, it is not unreasonable to say that as a global society we face some very substantial and complex immediate threats to the global social order as it exists today. But within these challenges also resides opportunity – humanity's big chance to get it right for future generations. In responding to the urgency and the promise of finally getting sustainable development right, the United Nations announced its long-anticipated new framework for global action. Launched in New York in September 2015, the 2030 *Transforming Our World* agenda promises to be the most comprehensive and inclusive effort to positively change the world in all of human history. The document was heralded at the time of its release as nothing less than a charter for people and the planet for the 21st century. While it remains to be seen whether it will meet these high expectations, there is no question that the *Transforming Our World* agenda raises the ceiling on sustainability globally. While it did not receive the same media attention as the later climate negotiations in Paris, the announcement of this agenda was at least as important, if only because it deals with the damage we are doing to other elements of the Earth system that is exacerbating and being exacerbated by climate change. *Transforming Our World* recognizes that unless we address the problems that form the backdrop to the climate threat, it will not be possible to prevent runaway changes in Earth system function that could bring to an end the conditions that make life possible on this planet as we know it today.

Transforming Our World is constructed around five themes: people, planet, prosperity, peace and partnership. It is also important to note that the agenda applies to the developed world just as much as to developing nations. In this context we may wish to remind ourselves of what "sustainable development" is commonly held to mean. The term refers to development that meets the needs of the present without compromising the ability of future generations to meet theirs. More specifically, sustainable development recognizes that eradicating poverty in all its forms, combatting inequality within and among countries, preserving the planet, creating sustainable and inclusive economic growth and fostering social stability are goals that are linked to one another and are interdependent. Our hope for achieving sustainable development globally resides in the balance among urgency, capacity and the will to succeed, as demonstrated by each and every UN member state in making action possible through common but differentiated responsibility at the level of each nation-state. It is at the national level that these goals must be met. The degree of our success will depend on governance, which in this context refers to the way in which authority is organized and executed in a society.

Transforming Our World not only demands far more focused national action within an enhanced framework of global co-operation and coordination, it also underscores the urgency of such action. The agenda makes it clear that we will not be able to deal with the degree of hydro-climatic change we are now witnessing on a global scale unless we are able to translate the 2030 sustainable development agenda into action at the national level. In other words, we won't achieve the goal of sustainable human existence at any meaningful level of prosperity unless we all take common global goals seriously and implement meaningful and measurable actions at the national level in every country in the world. This means there can be no laggards, particularly in the developed world. It also means the world cannot afford to leave anyone behind.

So the question becomes one of how any given country goes about integrating sustainable development goals into its national strategies. This will be particularly difficult in developing countries, where governments have limited means to act. It may be just as difficult in developed countries, however, where national governments have little interest in such strategies or choose to simply devolve responsibility for sustainability without attendant resources to sub-states or provinces, which in turn pass the burden down to cities, towns and rural districts. The 2030 *Transforming Our World* agenda holds that the manner in which we effect the translation from global threat to national action has to be seen as an urgent immediate priority for everyone fortunate enough to be able to give meaningful consideration to the future viability of their nation and the communities in which they live.

When compared to Europe, for example, North America is seen by the rest of the world as being a laggard in dealing seriously with matters of sustainability, especially as they relate to hydro-climatic change. *Transforming Our World* demands that Americans and Canadians alike ask themselves whether, as nations, they will just keep putting off meaningful action, leaving ever more complicated problems for those who follow after us to address, or instead will do something. At the time of this writing the most meaningful action – and in some regions of North America the only action – on climate change as it seriously relates to sustainability is being orchestrated by the mayors of cities. Big cities know how vulnerable they are to extreme weather events. They want change. They want resilience. They want sustainability. But they cannot achieve these goals on their own.

The fact is that North Americans can no longer ignore what is happening in the rest of the world. Nor can we deny our place in that world. We can't keep comparing ourselves to ourselves. We can't just keep saying it is too hard to change. The status quo can't and won't last. It has already been established that it is morally wrong for those alive today to claim there is nothing they can do because of their own incorrigible nature or circumstances. Our way of life in North America will change from within or be changed from without by emerging circumstances.

Often, in social movements, timing is everything. The world is about to reach out to steady itself. If we want change from within, now is the time. There is still room to move, but we have to move now while that room still exists.

One of the ways to re-energize the conversation about sustainable development and humanity's need for resilience in the face of rapid change is to talk about something none of us can live without: water. Of the nine critical Earth system boundaries that we dare not cross, water plays a significant role in seven. There are 17 goals set out in the 2030 *Transforming Our World* sustainable development agenda. While important in its own right, as treated in goal 6, water plays a role in 13 of the other 16 goals as well, which include ending poverty and hunger; ensuring healthy lives, inclusive education, gender equity and energy security; promoting sustainable economic growth, full employment, resilient infrastructure and sustainable cities and economies; and taking action on climate change.

The global goal with respect to water is to achieve universal and equitable access to safe and affordable water and sanitation for all. The goal also aims to improve water quality by reducing pollution, halving the proportion of untreated waste water and substantially increasing recycling and safe reuse globally. The *Transforming Our World* agenda also seeks to protect and restore water-related ecosystems in part by implementing integrated water resource management at all levels, including transboundary basins, by 2030.

The world learned from the earlier *Millennium Development Goals* that we need to better address the multiple roles water plays in establishing, maintaining and improving the human condition. We need to take advantage of the synergies that exist between effective water management and benefits that accrue not only directly to human health and well-being but also to the environment and the economy.

We are very good at evaluating the state of water

resource use, determining ecosystem health and evaluating potential climate impacts, but we are less capable of changing our practices once we have characterized those parameters. Broader, nexus thinking is critical, in part because normal professional thinking is narrowly focused on economic interests. This thinking is reinforced within institutional silos that arbitrarily separate needs and habits of utilization. Creating a systems approach to managing water has to be seen as synonymous with sustainability and resilience.

Cities

The 2030 *Transforming Our World* agenda focuses considerable attention on the not insignificant matter of cities. Some 92 per cent of the population growth that has brought the most recent 1.2 billion people into the world has occurred in cities. Some 60 per cent of the urban space required to accommodate future populations has yet to be built. Sustainable development goal 11 aims to make the world's cities and human settlements inclusive, safe, resilient and sustainable.

If we are to live sustainably on this planet, cities must commit to achieving the goals of the 2030 *Transforming Our World* agenda. This means that resilience has to be seen as a child of sustainable development. While the agenda does not provide enough guidance to help any given city in terms of specific pathways to resilience, it does provide clear recognition of the critical importance of municipalities with respect to the goal of creating a sustainable civilization. While cities can certainly act in any manner consistent with their own local needs and vision with respect to resilience, what they do must contribute to the global sustainability effort. Sustainable development in cities means adequate, safe and affordable housing and basic services for all, as well as safe, affordable transportation systems. It also means strengthening and safeguarding cultural diversity.

The *Transforming Our World* target of providing universal access to safe, inclusive and accessible green and public spaces is, of course, consistent with all efforts to improve resilience, especially in cities vulnerable to heat island effects. Rising levels of urban heat are now seen to constitute the single greatest climate-related threat to human health globally. Urban heat waves now account for more deaths per year globally than all other forms of extreme weather. We don't have to wait for the future for this. It's happening now. The *Transforming Our*

World vision is that by 2030 sustainable, resilient cities will have significantly reduced the number of deaths and the economic and psychological effects caused by disasters, including water-related catastrophes. The target is that as early as 2020 we will have substantially increased the number of cities and other human settlements that are implementing plans to mitigate and adapt to climate change and enhance resilience to disasters. The template for holistic disaster risk management at all levels is the United Nations-supported Sendai Framework for Disaster Risk Reduction.

Urban regions

By 2030 we also want to reduce the adverse per capita impact of cities on surrounding regions. This target, of course, speaks directly to air quality, water contamination and other waste management issues. But the agenda does not stop there. Another global sustainable development target that has great relevance here is recognition of the need to support positive economic, social and environmental links between urban, peri-urban and rural areas through the strengthening of national and regional planning.

It is time, in North America, to start examining just how much damage we are doing outside our cities in order to sustain urban viability and prosperity. Those who spend any time at all travelling in rural areas often claim with considerable confidence that most city-dwellers would be very surprised at what is going on out there in the name of urban prosperity. You don't have to go very far outside many major North American cities before you get the sense that anything that has any monetary value at all is being cut down, ripped up or sucked from the ground and hauled away in what appears to be a mad scramble for wealth. Despite claims to the contrary, much is happening that could never pass for sustainability.

Given that North America once had one of the best reputations in the world for responsible environmental management, one might ask how we could have gone so far in the other direction in such a short time. A partial answer to that is the movement away from the principle of environmental protection. In the heyday of North America's international reputation for leading-edge environmental governance, jurisdictions at all levels were driven by the desire to preserve natural places and protect natural system function. Regulations forbidding pollution and other forms of environmental damage were not considered restraints as they are today but as protections.

In the 1980s, however, the principle of environmental protection was superseded by the notion of environmental "assurance." The objective became to mitigate impacts rather than prevent them. And we created markets and industries that profit from doing this. It did not take long for it to become conventional wisdom that the effects of any kind of development could be mitigated, regardless of scale. Consequently, North America is now dotted with environmental damage "super-sites" and abandoned or orphaned resource developments that will require expensive attention to avoid serious polluting effects for which the public will have to pay for centuries to remedy. In some particularly dangerous cases, such as the Giant Mine site in northern Canada, we will have to pay for the stabilization of contaminants for as long as we occupy this planet.

The 2030 *Transforming Our World* global agenda makes it very clear, however, that sustainable development can no longer simply aim for environmentally neutral solutions. If we are to achieve any meaningful level of sustainability, all development has to be not only sustainable but restorative. We can no longer simply aim to slow or stop damage to the Earth system; we have to thoughtfully restore declining Earth system function. A good first step

would be to create a global business model that respects the real value of ecosystem services rather
than simply creating a market for repairing the uncalculated and often incalculable damage we do to
those services as a matter of prescribed course.

Agricultural landscapes

Many North Americans now subscribe to the principle put forward by Wes Jackson while he was still
active with the Land Institute in the US that for any
given landscape in human use there should be a high
enough "eyes to acres" ratio to save it from destruction. By "eyes" Jackson means competent watchful-
ness that is knowledgeable about the history and
rhythms of place, a constant presence that is always
alert for signs of harm and also of health. If you
don't believe such damage occurs, I would invite you
to go see what is happening in your state or province
in the name of concentrated urban wealth. In the interests of full disclosure with respect to urban resilience, the question that might be asked if you live in
a city is, how much resilience do places outside your
city have to lose to ensure yours?

Nowhere is this truer than in agriculture. As has
been noted by writers such as Wendell Berry, over
the past 50 years farming has come to be controlled

by the demands of machinery rather than the nature of agricultural lands. Universities, corporations and governments continue to almost unanimously support industrial agriculture despite the now almost overwhelming evidence of its negative impacts. These effects include soil erosion and salinization, aquifer depletion and dependence on fossil fuels and toxic chemicals. The damage caused by industrial agricultural practices also includes pollution of streams, rivers and lakes; loss of genetic and biological diversity; and the destruction of rural communities and cultures of animal husbandry that follow in its wake. It is not too much to claim that we are now in a situation in North America where we face the real threat of creating a cannibalistic economy in which one sector thrives only at the ultimate expense of others.

While industrial agriculture appears to be undermining itself, it also faces a very significant existential threat. We now realize that climate change has the potential to irreversibly damage the fundamental resource base upon which agriculture depends, with grave consequences for food security globally. Even though land abuse, soil loss and desertification are among the greatest threats to the future of humanity, there is a taboo against criticizing agriculture.

It is widely held outside of North America that agriculture is in a state of emergency that cannot be sustained indefinitely. The concern is that many of our industrial farming practices may be self-terminating. That should matter to all of us. Why can't we talk about these things? Why are we so afraid of disagreeing with another? We know how important agriculture is. We all need food; nobody disputes that. But we also need water and energy. If we want resilience in our cities – or anywhere else – we need changes not only in agricultural practices but in agricultural principles.

What we really need is another agricultural revolution. We need to focus on critical interdependencies, especially as they relate to water, food and energy. Secure and reliable access to water is a necessary condition for food security. One of the *Transforming Our World* goals pertaining to ending hunger that relates directly to the food-growing regions of North America is to implement agricultural practices that increase productivity and production while at the same time helping to maintain ecosystems, strengthen capacity for adaptation to climate change and reduce the impacts of extreme weather, drought, flooding and other disasters that progressively diminish land and soil health. There is no time to lose. The goal is to do this by 2030.

For example, we will have to cut per capita global food waste by half at the retail and consumer level and reduce food losses all along production and supply chains, including post-harvest losses. In fact, we will have to substantially reduce waste all across the board.

The year 2030 is also the deadline for the world to achieve sustainability in managing and efficiently using natural resources. We have to figure out how to manage chemicals and reclaim wastes throughout their life cycles.

People everywhere have to have the relevant information and awareness of what they can do in terms of their own lifestyles to help themselves and their communities achieve sustainability. We have to make the transition from seeing waste as waste to seeing waste as wealth. If we don't want these kinds of problems to get away on us, we have just 15 years to do this.

Regrettably, the fact that is emerging is that the bulk of the goals in the 2030 *Transforming Our World* agenda will be difficult or impossible to meet unless we address the climate threat first. If we are to have any hope of achieving meaningful sustainability globally, we have no choice but to combat climate disruption and its effects. We can protect ourselves from climate impacts by improving public education, enhancing awareness of the issues as they develop and increasing institutional capacity directed at mitigation, adaptation, impact reduction and early warning. To achieve this end, every country in the world has to integrate climate change into its national policies, strategies and planning. We could start down that road by rationalizing markets to remove the distortions caused by inefficient fossil-fuel subsidies that encourage wasteful consumption.

RANKING THE RISKS

It is important to be very clear on this point: climate impacts will affect the development trajectory of all nations, rich and poor. In fact, there is a proven link between climate disruption and dedevelopment. While the link between recurring extreme weather events and the challenges of maintaining critical physical and social infrastructure in developing countries has been noted by organizations like the United Nations and the World Bank, the extent to which climate disruption has begun not just to slow but to reverse economic development was not widely recognized as a global economic threat until 2016. In January of that year, the World Economic Forum, which annually presents its ongoing research findings at a high-profile policy conference in Davos, Switzerland, released its 11th *Global Risks Report*. As part of that survey, nearly 750 experts assessed 29 separate global risks for both likelihood and possible effect over a ten-year time horizon. The current risk with the greatest potential impact in 2016 was found to be the failure of climate change mitigation and adaptation. It is important to note that this was the first time since the forum's original annual report was published in 2006 that an environmental risk had the top ranking. The failure to mitigate and adapt to climate change was now considered to have created greater potential risk of damage than the next four risks, in descending order: weapons of mass destruction; water crises; large-scale involuntary migration and the economic consequences of energy price shocks (whether increases or decreases).

When the risks were ordered in terms of likelihood alone, the most serious one was thought to be large-scale involuntary migration, followed in descending order by extreme weather events; failure of climate change mitigation and adaptation; interstate conflict with regional consequences and major natural catastrophes. It was also noted that such a broad risk landscape was unprecedented in the history of the report series. The authors also pointed out that this more diverse array of risks comes at a time when the toll from global risks appears to be on the rise.

The report noted, for example, that the global climate warmed enough in 2015 to raise the Earth's average surface temperature to the milestone of 1°C above the pre-industrial era for the first time in human history. The forum also asserted that according to the UN's Refugee Agency, the number of people around the world forcibly displaced in 2014 stood at 59.5 million, and that this number was nearly 50 per cent greater than the 40 million refugees that were on the move at the time of the Second World War. According to the UN, one in four of the world's refugees is now Syrian, with 95 per cent relocated in surrounding countries. The report also noted that available data supported the likelihood of increasing risk annually in all the categories the World Economic Forum had been measuring since 2014.

In addition to assessing likelihood and potential impact, the *Global Risks Report 2016* also explored interconnections among risks. The report estimated that a very small number of risks accounted for interconnections compared with 2015. The two most interconnected risks in 2015 were profound social instability and structural unemployment or underemployment. Margareta Drzeniek-Hanouz, the head of the global competitiveness and risks program at WEF, underscored the importance to leaders of understanding such connections as a means of prioritizing areas for action as well as planning for contingencies: "We know climate change is exacerbating other risks such as migration and security, but these are by no means the only interconnections that are rapidly evolving to impact societies, often in unpredictable ways. Mitigation measures against such risks are important, but adaptation is vital."

In reading this WEF report one could easily get the impression that in terms of risk it is difficult to choose which panic button to push first. The circumstances as they stood in 2016, at least in terms of impacts on global economic performance, were probably explained best by Cecilia Reyes, the chief risk officer for the massive Zurich Insurance Group:

> Climate change is exacerbating more risks than ever in terms of water crises, food shortages, constrained economic growth, weaker

societal cohesion and increased security risks. Meanwhile geopolitical instability is exposing businesses to cancelled projects, revoked licences, interrupted production, damaged assets and restricted movement of funds across borders. These political conflicts are in turn making the challenge of climate change all the more insurmountable – reducing potential for political co-operation as well as diverting resources, innovation and time away from climate change resilience and prevention.

While not appearing in the WEF report, a haunting image of the extent and nature of contemporary risks was put forward at the 2016 forum by another global insurance giant, Munich Re. The image was a map of interconnections between various economic, environmental, geopolitical, societal and technological risks associated with the failure to effectively and meaningfully adapt to climate change. What the map illustrated was the cascading effect of the failure to adapt to hydro-climatic change. On a global scale, failure leads first to greater vulnerability to extreme weather events, food crises, water crises, large-scale involuntary migration and further man-made environmental catastrophes, which in turn lead to biodiversity loss and Earth system collapse.

This wasn't speculation; it was already happening. The terrible violence rocking Syria and the spillover effects in Europe in 2015 did not start as a war; it all began with a five-year drought that contributed to sparking a war. What we haven't understood until now is the extent to which the fundamental stability of our political structures and global economy is predicated on relative hydrologic predictability. As a result of the loss of relative hydrologic stability, both political stability and the stability of our global economy in a number of regions in the world are now at risk. We are only now beginning to understand how complex this issue has become.

What if the damage doesn't stop? To answer that question we should be paying attention to what happened and is happening to California. At the time of this writing, California appeared to have been dealt a hydro-climatic double whammy: a cyclical return of drier conditions, combined with the effects of warming. There was drought from Mexico to Alaska and from the Pacific Coast to thousands of kilometres inland. And, of course, there also is a direct link between drought and fire. Conversely, other parts of California were hit by major flooding that was caused in part by a super El Niño forming in the Pacific. It's not that it isn't over yet – it isn't ever going to be over. It is now apparent that the hydro-climatic conditions most of us in North America grew up with and have become accustomed to may not be with us much longer, nor will they return during the lifetime of anyone alive today.

There is not much of a silver lining in all this. The non-monetary cost of climate disruption is migration. We know it's going to happen, so what are we going to do about it? One thing we can do is realize that in times of climate disruption, smart people and smart money move, which is exactly what we are seeing now in California. As many experts have observed, municipalities with foresight might be able to attract mobile businesses ready to flee the sites of future climate-related disasters by creating conditions favourable to relocation. To be one of the places money moves to, however, you have to have enough water and be able to catch up with and get ahead of your own sustainability challenges. In his 2013 book, *American Exodus*, Giles Slade cited research that predicted that as many as 600,000 Americans may be moving north into Canada to escape changing climatic conditions in the next ten years. That prediction is not likely to sit well with many Americans, or for that matter with every Canadian.

ACHIEVING TRUE SUSTAINABILITY

Even though it often sucks all the air out of a room to talk about it, climate disruption is only one of the 17 global sustainability challenges we need to address through the 2030 *Transforming Our World* sustainable development agenda. We have to better protect our oceans, first from land-based activities, including marine debris and nutrient pollution. We also must protect and restore coastal ecosystems, regulate marine harvesting and end overfishing, and halt perverse subsidies and destructive fishing practices. There is urgency in this. If we don't minimize and reverse the impacts of ocean acidification by 2030, we will lose one of humanity's most important sources of food and livelihood.

To achieve these goals we must increase scientific knowledge, further develop research capacity and stimulate faster transfer of marine knowledge and technology.

The UN has also set 2030 as the goal for combatting desertification, restoring degraded land and soil, halting the degradation of natural habitats, minimizing the impacts of invasive species and halting the loss of natural biodiversity, with the aim of ensuring that sustainable development remains possible in the future.

We cannot achieve these goals without making the world a safer place. In order to avoid slowing or even reversing sustainable development, we have to significantly reduce all forms of violence and related deaths everywhere. We must, all of us, promote the rule of law at all levels and reduce corruption and bribery in all their forms.

Our sustainability also depends on reducing illicit flows of money and arms and on combatting all forms of organized crime. We must demand responsible, inclusive, transparent, participatory and representative decision making wherever we live.

Looking at this long list of goals, it becomes very clear that accomplishing the *Transforming Our World* agenda will require additional financial resources, especially if these goals and targets are to be met by 2030. The question then becomes whether or not there is enough money in all the world to achieve so many goals and targets in such a short time. Economists believe the money is there but not in the traditional places. It may not be easy in many cases to free it up, but it appears there is enough public and private wealth available to transform our world in ways that will make a sustainable future possible.

The first step in financing these goals is to have developed countries fully honour their official development commitments. If, as a nation, you promise to help the rest of the world with money or technology, you should meet that promise with no unreasonable strings attached.

We will also have to find new ways to pay for what we need to do. Subject to rigorous due diligence, national governments must identify, explore and utilize new and emerging financial resources. It will be critical to encourage and promote effective public, public–private and civil-society partnerships in efforts to develop new kinds of financing and resourcing strategies.

All 17 of the sustainable development goals and all 169 of the targets in the 2030 *Transforming Our World* agenda are held to be universal, indivisible and interlinked, and as such they should all be regarded as having equal importance and be accorded equal priority for implementation. While the agenda respects each country's policy space, it is recognized that national development efforts need to be supported by an enabling international economic environment. That environment must include coherent and mutually supportive world trade and monetary and financial systems, as well as strengthened and enhanced global economic governance.

The very real spectre of unanticipated new issues emerging between now and 2030 is virtually a given. We will need to expand the data available upon which to make sustainable development decisions to include broader, satellite-based, Earth system observations and geospatial information.

Implementation of the *Transforming Our World* agenda will be tracked to make sure no one is left behind. The goals and targets will be followed and reviewed using a set of global indicators that were scheduled to be put in place by March of 2016. All reviews of progress toward implementation of the agenda must be country-led and country-driven. They should be periodic and inclusive and they should draw on observations and contributions by Indigenous peoples, civil society, the private sector and other stakeholders. Ideally, such reviews should provide a new platform for partnerships.

There is a lot to do. Is there economic opportunity in pursuing these goals and targets? Absolutely. If we want to have meaningful and prosperous lives in the coming decades, achieving these goals now must become a pillar of every economy, nationally and globally. Theoretically, all the elements required to create sustainability are included in the agenda. The great challenge and urgency is to make these goals and targets a priority at the national level. To this end, all member states are encouraged to develop ambitious national responses related to the implementation of the *Transforming Our World* agenda as soon as practicable.. This does not by any means suggest starting over. What it means is building on and focusing existing planning instruments and strategies for sustainable development and resilience enhancement. As people, we depend on the planet for prosperity, which can only be sustained through peace and partnership. If North America gets behind this agenda and we all work together globally, we can transform the world for the better by 2030.

CHAPTER 2

Why a Stable Climate Is Critical to Sustainability

Because I live in the Rocky Mountains in Canada, I spend a lot of time paying attention to changing weather. It often strikes me that the people where I live – like everyone else in Canada and the US – also listen to local weather forecasts and judge the predictions based on their own particular interests and immediate plans. We joke about the local foibles of weather because we are so familiar with what we know about what weather might do to our expectations.

But what if the local forecasts in the Rockies were to start reporting weather we had never experienced before? How would we like 45°C daytime temperatures, cooling to only 35° at night? What if the forecast, instead of calling for afternoon showers, told us to be prepared for relentless rainfall that will last for a week without a break? Or worse yet, what if suddenly it never rained at all? What if forecasts called for tornadoes in places they never occurred before, or hailstones the size of baseballs, or three metres of snow in two days? We keep thinking linearly. We still believe our climate will be more or less as it is now, if perhaps a little warmer. But that's not what is going to happen. Weather doesn't work that way. When climate changes, it changes in steps. Each progressive step can be different from the last – more energetic, more violent, more fatal. Unless we stop altering the composition of our atmosphere, our weather will continue to change until eventually we won't be able to stand it anymore. We are living now in a bubble, and when that bubble breaks we are going to witness a world of change. That is what the UN Intergovernmental Panel on Climate Change and, in the United States, the National Oceanographic and Atmospheric Administration and NASA would have us prevent.

WATER SECURITY
The more we focus on climate change, the more the emphasis is on water and in particular on water security. Water security used to mean being able to reliably provide adequate water of the right quality

where and when you need it for all purposes, especially agriculture but also for purposes related to sustainable, natural, biodiversity-based Earth system function. It also used to mean ensuring that your use and management of water in the region where you live did not in any way negatively affect the water security of regions up- or downstream from you, currently or in the future. Water security still means all of these things, but changing circumstances now include an additional element of water security that must be considered.

Over the last decade water security has also come to mean being able to achieve these goals in the face of not just growing populations but also new circumstances created by the acceleration of the global hydrological cycle. The era we live in might be called the storm after the calm. After a period of relative hydroclimatic stability, during which we created most of our built environment, step-like changes to our hydroclimatic circumstances are demanding that we redefine what development and sustainability mean, not just nationally but globally. This, in turn, demands that we reassess personal and collective vulnerability, accountability and liability and adapt quickly to changed circumstances if we want to sustain our prosperity in the face of altered hydro-climatic conditions.

What such reassessment reveals is that security of water, food and climate, respectively, are inseparable; each one is implicit in the others. It could even be said they are the same thing. Water, food and climate security are critical elements of sustainability. Without stable water and climate regimes, sustainability will forever remain a moving target. But flood resilience is also very much an element of the larger water security ideal. This makes forest management, especially in upland regions, a critical factor in any water and climate security formula. These are old ideas made new again in the context of the emerging politics of hydro-meteorological change. In order to see where all this might be taking us, it may be helpful to examine how these politics of hydro-meteorological change emerged and have evolved.

FLOODING

While not related directly to climate change, the issue of urban flood resilience appeared on the radar of research networks just after the UN Water for Life Decade was initiated ten years ago. What happened in New Orleans in 2005 could simply not be ignored. While the media and most public attention focused on the failure of the US Federal Emergency Management Agency and the culpability of the US Army

Corps of Engineers in the disaster, the UN Water for Life Decade partnership in Canada looked at the broader significance of the event. We examined the implications of a projected increase in the vulnerability of big cities to longer, more frequent and ever more intense flooding events that were expected as a consequence of human-caused changes in the composition of the global atmosphere. What we found was that, while the initial focus was on the huge cost of repairing the damage to the city, the real cost – the deeper cost that went largely uncalculated – was the permanent physical and psychological impact on those who survived Hurricane Katrina and its aftermath. It has taken a decade to sort out just how serious that damage really was and remains.

Anyone interested in the larger issues related to the sustainability of our society in a changing climate who hasn't already read Sheri Fink's book, *Five Days at Memorial: Life and Death in a Storm-Ravaged Hospital,* is urged to do so. In the aftermath of Katrina, sections of New Orleans were uninhabitable for weeks. The medical centre where Fink worked was an island in the middle of the flood zone. When the power went out in the city, the hospital's backup generators could not keep the air conditioning on and still supply light. Helicopters could only take one or two at a time of the 2,000 people that needed to be evacuated. The ethical question became whom to evacuate first. How do you prioritize who lives and who dies? In the aftermath, some doctors and nurses were charged with murder. What the episode demonstrated is that the moral jeopardy that arises in the aftermath of extreme weather is similar to what happens in war zones, where it becomes impossible to adhere to established moral values. That is the larger terrain we are entering with respect to extreme events.

Katrina was followed by nearly a decade of foreshadowing of the flood disaster we later experienced in southern Alberta. There had been flooding there in 2005, followed by widespread inundation throughout Europe and the northern hemisphere almost every year afterward.

Then, in 2010, we began to see mega-floods, events in Australia and Pakistan so large they had never been experienced before. Then there was a mega-flood on the Canadian prairies in 2011. It was clear there was something going on out there – the hydrologic order was changing – but we didn't have the evidence to prove it. Then suddenly we did have it, almost simultaneously in both Canada and the United States.

HYDROLOGICAL STATIONARITY
AND EARTH SYSTEM FUNCTION

In the fall of 2011, John Pomeroy and researchers at the University of Saskatchewan showed evidence that was confirmed by a major report released at the same time by the US National Research Council that indicated that the global hydrological cycle is, in fact, accelerating. The report confirmed how serious the loss of hydrologic stability could be in North America and around the world if current trends persist. The findings of the National Academies analysis include consensus on the fact that anthropogenic land cover changes such as deforestation, wetland destruction, urban expansion, dams, irrigation projects and other water diversions have significant impact on the duration and intensity of floods and drought.

The report concluded that "continuing to use the assumption of stationarity in designing water management systems is no longer practical or defensible." In other words, the old math and the old methods no longer work, and continuing to use them will in time be technically if not legally indefensible.

The significance of the loss of hydrologic stationarity is slowly beginning to sink in. At present, however, action in support of true sustainability and resilience in the face of hydro-climatic change in North America appears to be moving along at five kilometres an hour while the problem is moving along at 19 kilometres an hour and accelerating. We need to catch up while we still can.

Because of the increasing number and growing costs of climate-related disasters, more and more people in North America are becoming concerned about resilience. This growing interest coincides with a critical time in the global dialogue concerning the sustainability of human presence on this planet.

Those who remain skeptical about whether or not climate change is real or who believe it isn't necessarily a bad thing are urged to consider other impacts we are collectively having on the Earth system, the effects of which are coming at us fast. We have physically altered the character of 60 per cent of the Earth's surface and are well on the way to wiping out as many as half of the rest of the life forms that share this planet with us and are big enough to see with the naked eye. We have changed the chemistry of the global ocean, accelerated the rate and manner in which water moves through the global hydrological cycle and disturbed natural precipitation patterns. And now, on top of all of this, we are changing our climate. This means we have entered a new epoch in which we can no longer count on self-willed, self-regulated natural landscapes to absorb human impacts on Earth system function.

Whether we like it or not, we have to assume responsibility for staying within Earth system boundaries. This means we have to rethink sustainability. Despite inherent tensions among them, the next iteration of global sustainable development goals and targets must create a safe operating space within Earth system and social boundaries. The 17 goals and 169 targets set out in the UN's 2030 *Transforming Our World* global sustainable development agenda aim to do just that. But just as the World Economic Forum's *Global Risks Report 2016* demonstrated important interconnections among the various threats to global economic stability, there are also interconnections among the goals of the 2030 *Transforming Our World* sustainable development agenda. While they all must be linked together, some sustainable development goals are prerequisites for achieving others. Climate action is one of those goals that, if it is not achieved, will make it difficult if not impossible to achieve many of the others.

What elevates the issue of climate action to its current and growing importance is that we cannot achieve success in addressing the 16 other global sustainable development challenges, which include huge problems such as eliminating poverty and hunger, unless we stabilize the composition of the Earth's atmosphere. To do that, we have to stop filling the sky with our unwanted greenhouse gas emissions.

The composition of the atmosphere is the linchpin holding the ice–water–weather–climate system of the planet together. If we cannot stabilize the Earth's atmosphere, we cannot know which way our sustainability efforts must trend, because we will not know the conditions to which we will need to adapt.

Because of warming's effect on the global hydrologic cycle, we have to deal with the instability of the atmosphere or we will not be able to achieve the goal of providing clean water and sanitation for all.

Without stabilizing the composition of the Earth's atmosphere, we will no longer be able to reliably predict where our food will come from. Droughts, such as those that occurred in California recently and in Syria between 2006 and 2011, are beginning to happen more widely and will continue to do so for the foreseeable future, threatening food security everywhere. But we don't just grow food on land; without stabilizing the Earth's atmosphere we cannot reverse the acidification that is threatening ocean food webs. What we learn from this is that we cannot have food security without climate security, and without food and climate security there cannot be peace and justice for all.

But there is more. Our cities were designed for climatic circumstances that will soon no longer prevail. Without stabilizing the composition of our atmosphere, we will not know to what standards we need to redesign our cities in the face of ever more powerful storms, bigger floods and longer-lasting heat waves. The complicated challenge of accurate prediction aside, we cannot know how much sea level rise we need to protect our cities against. Without stabilizing the composition of the atmosphere, urban resilience will remain a moving target, constantly receding beyond our grasp.

Nor can we, without stabilizing the atmosphere, prevent desertification, halt ecosystem collapse or slow the accelerating rate of extinction of the other creatures with which we share this planet.

It is now very clear that the relative climatic stability we have enjoyed over the past century or so will not be returning for centuries, if ever. Failure to realize this could cost us our prosperity. It could even cost some of us our lives.

It is clear also that we need to define a safe place in terms of sustainability toward which all of humanity must aim. In this quest, knowledge is not enough; we need the will and the permission to solve global change problems. That is why COP21 in Paris in December 2015 was so important, because, as the motivating slogan of the conference put it, "Later, it will be too late." How we did in Paris, then, is critical to our future.

CHAPTER 3

River of No Return:
COP21 in Paris and Our Climate Future

River of No Return was a 1954 American Western film directed by Otto Preminger. Set in the northwestern United States in 1875, the film focuses on a widower named Matt Calder, played by Robert Mitchum, who recently has been released from prison after serving time for killing a man while defending another one.

Calder arrives in a boomtown tent city in search of his 10-year-old son Mark, who was left in the care of a dancehall singer named Kay, played by Marilyn Monroe. Kay's fiancé, gambler Harry Weston, played by Rory Calhoun, tells Kay they must go to Council City to file the deed on a gold mine he won in a poker game. They head downriver on their flimsy log raft, and when they encounter trouble in a series of rapids, Matt and Mark rescue them. Weston offers to buy Matt's rifle and horse so as to reach Council City by land, and when Matt refuses, Harry knocks Matt unconscious and steals both. Kay chooses to stay behind to take care of Matt and Mark, and the three are stranded in the wilderness. When hostile local Natives threaten, the three are forced to escape down the river on Harry's rickety raft. Their lives are changed utterly by the rapids they encounter.

All of the river scenes in *River of No Return* were shot in Banff National Park in the Canadian Rockies. It is a wilderness that in many ways is easy to compare to the one the world uncertainly entered in Paris with respect to climate change.

What we have embarked upon in having at last woken up to the threat of climate disruption is nothing less than a ride on a flimsy log raft down a river of no return. The whole world is on this raft and we are heading for the rapids fast. Like the wilderness situation that Matt, Mark and Kay found themselves in, the rapids are only part of a larger setting in which humanity finds itself dangerously situated. Similarly, climate change is just one of many hazards we face.

In September of 2015, just two months before the

Paris climate talks, the United Nations responded to the urgency and the opportunity of finally getting sustainable development right. Its response was the announcement of the 2030 *Transforming Our World* agenda. The scale of the *Transforming Our World* challenge should not be underestimated. The UN's global sustainable development agenda is about getting ourselves out of a very dangerous situation we as a civilization have gotten ourselves into. As noted, there are 17 goals in the 2030 *Transforming Our World* agenda. Serious as it is, climate action is only one of them, though a critical one. The *Transforming Our World* agenda relies very heavily on how quickly and effectively the global community can respond to the climate change threat. The outcomes of what transpired in Paris in late November and early December of 2015 will determine whether humanity will have a sustainable presence on this planet. So what, in fact, happened at that conference?

SEPARATING THE HYPE FROM THE HOPE IN PARIS

The hype
Perhaps it was British journalist George Monbiot who said it best. By comparison to what it could

have been, what happened in Paris was a miracle. Compared to what it should have been, however, it was a disaster. As American climate change action advocate Bill McKibben pointed out, we are no longer just sitting around the table negotiating with other countries. We are dealing with fundamental atmospheric physics, and the physics holds all the best cards. It means that if average temperatures on Earth rise by 4°C to 5°C, significant areas of the planet are likely to become uninhabitable for at least parts of every year.

The Paris agreement continues to represent a linear approach to the problem of leaving all of humanity vulnerable to unanticipated step-like changes in our climate-related circumstances.

One does not have to read far into the agreement to realize we have wasted a generation in responding to the climate threat. There are many things in the Paris agreement we should already have accomplished, such as meaningfully reducing greenhouse gas emissions, developing methodologies for assessing adaptation needs with a view to assisting developing countries; strengthening regional co-operation on adaptation; climate-proofing national and regional economies; and developing integrated approaches to averting, minimizing and otherwise addressing large-scale displacement of people as a consequence of climate disruption.

The Paris agreement allows for examination of the risks of damage and displacement but does not allow attribution of blame, suggestions of liability or recommendations for compensation. If you are an island state about to be submerged, there is no one you can hold directly accountable.

Because it has taken us 20 years just to agree on goals, pace has now been identified as everything. That said, the Paris agreement grants the world four more years from the time of this writing, not to set the pace as might be expected but just to set national emissions reductions targets. Prior to 2020 nothing is obligatory.

The degree to which this agreement is largely aspirational is obvious in that it calls for the establishment of a mechanism for tracking emissions at a national level but maintains that even these contributions to global reduction targets are still voluntary. The agreement "recognizes" the important role of providing incentives for emissions reduction activities, including tools such as domestic policies and carbon pricing, but that's all.

The agreement rightly suggests that governments cannot and should not be allowed to address the climate threat by themselves. The agreement only "welcomes the involvement of the private sector, civil society, financial institutions, cities and other sub-national actors to join" governments at all levels in addressing the climate threat. That the agreement did not say it "expects" the participation of these entities or even "demands" participation of these entities in implementing solutions only underscores its aspirational character.

Then there is the not insignificant issue of time frames and hard deadlines. The first global stocktaking of implementation progress will not take place until 2023. Even participation in mechanisms for establishing targets for greenhouse gas emissions reductions beyond 2023, to mid-century and later, is voluntary. In a very real sense these conditions in themselves undermine the agreement from its very inception. In the five years the parties to the convention have granted themselves to establish how much they will contribute to global emissions reductions, we could very well blast past any real opportunity to limit mean warming to 1.5°C. The delay in action could even put the 2°C target out of reach. All self-congratulations aside, it appears that once again all we have done is kick the can down the road. Cut the numbers any way you want, but at the end of the

Paris conference we still have less than a 50 per cent chance of avoiding runaway climate impacts.

The Paris agreement also has other shortcomings. There is no mention of direct human health risks. Aviation and shipping are not included. Water security is not mentioned, even though effective management of water is the very foundation of climate security. The agreement focuses mostly on technology and technological transfer while largely ignoring the improvement of ecosystem function in service of keeping the world from warming more than 1.5°C.

While the agreement does recognize the critical need to reduce emissions generated by deforestation and forest degradation, there is no reference beyond forestry management to ecosystem-based mitigation and adaptation strategies such as enhancing soil health as a means of increasing carbon sequestration. In this agreement, agriculture receives a "get out of jail free" card.

Nor is it clear where the money is going to come from to finance even what has been proposed in terms of climate action. Though much is made in the agreement about transparency, it nevertheless allows developing countries to weasel out of reporting by making such reporting at the level of in-country reviews optional.

The agreement relies on transparent, non-adversarial, non-punitive compliance. It is also an agreement that is easy to get out of. At any time after three years from the date the agreement comes into force for a given signatory, the signatory can withdraw from the process by simply giving written notification of its intention to do so. The withdrawal will take effect one year after the notification is filed.

There is also some question as to whether this agreement can withstand outside disruptions such as large-scale terrorism events and cyclical economic collapses that neither economists nor politicians appear to be able to predict or control. Because it is non-binding, the agreement is also highly vulnerable to political manipulation. The question then becomes whether it can survive political turmoil or the election of radically different new governments. How would the agreement fare, for example, if yet another neoliberal – someone like Donald Trump – were to be elected president of the United States?

The biggest failing of the Paris agreement, however, may reside in the fact that it offers only a long list of urgings, invitations and encouragements to signatories to act now and in the future. There is nothing to force them to do so. At this time, the agreement merely "urges" parties to make voluntary contributions to the timely implementation of the process to which they agreed. At the time of this writing, we have not yet reached the peak of greenhouse emissions, while the Paris agreement only promises that we will eventually begin to reduce them. Knowing this, it is hard to be hopeful. But there is hope.

The hope

The real miracle in Paris was that, for a moment at least, we got 193 nations – rich and poor – to agree on something at the same time, however limited the consensus may be. That is an achievement in itself. Whether everyone will continue to agree once the delegations return home and get dissuaded from action by their fractious political constituencies remains to be seen.

No, it's not perfect – in fact, it is far from perfect – but now at least we have something to build on – something concrete we can work together toward that didn't exist until COP21. This is a beginning, not an end.

The first thing that is really important about the Paris agreement is that it nests climate action within the larger context of the UN's 2030 *Transforming Our World* sustainable development agenda, as well as other important UN conventions.

It has often been said that addressing the global climate threat will require nothing less than the kind of vision, concentration of finances, resources, intelligence and purpose that permitted the United States to put a human on the moon in 1969. In nesting the climate challenge within the expanded global dialogue concerning the sustainability of human presence on this planet, this agreement recognizes that what we have arrived at is the need for multiple moon-shots, in each of the 17 areas critical to sustainability, and that we have to urgently embark upon all of these simultaneously.

In this context, the agreement clearly recognizes that climate change represents an urgent and potentially irreversible threat to human societies and the planet and thus requires the widest possible co-operation by all countries, and their participation in an effective and appropriate international response, with a view to accelerating the reduction of greenhouse gas emissions. It also acknowledges that deep reductions in global emissions will be required in order to achieve the ultimate objectives of avoiding dangerous anthropogenic warming.

The agreement importantly recognizes the need for an effective and progressive response to the urgent threat of climate change on the basis of the best available scientific knowledge. The COP21 agreement makes it very clear that, no matter how difficult it will be, the global effort must be to hold the increase of global average to well below 2°C, and to 1.5° if possible.

In establishing the 1.5° target, the agreement recognizes the specific needs and circumstances of developing countries and those particularly vulnerable to harmful climate effects. Conversely, the agreement also acknowledges that many nations may be affected not only by climate change but also by the impacts of measures taken in response to it. The agreement clearly sets out the fundamental priority of safeguarding food security and ending hunger, and takes notice of the particular vulnerabilities of food production systems to the adverse effects of climate change.

While acknowledging that climate change is a concern common to all of humankind, the Paris agreement also makes it clear that actions taken to minimize its effects must be respectful of human rights, the right to health and the rights of Indigenous peoples, as well as gender equality and intergenerational equity.

Within all these parameters, the agreement calls for the setting and achieving of economy-wide absolute emission reduction targets at the national level. The signing parties have to formally submit their targets for emissions reductions no later than 2020 and resubmit revised targets every five years thereafter. Signatories are bound in such submissions to clearly and transparently include common baseline references, such as the year to which emissions reductions are being compared; methods utilized in estimating and accounting for anthropogenic emissions; assumptions and methodological approaches in arriving at targets; timelines for implementation, and explanation of how the particular signatory's reductions of emissions contribute to the objective of strengthening the overall global response to the climate change threat. The common methodologies for accounting will be established by the Intergovernmental Panel on Climate Change and will require that parties do not double count or arbitrarily fail to include carbon sinks or sources.

Article 5 of the agreement recognizes the importance of preserving and enhancing carbon sinks and provides clear marching orders in this regard for forestry management. The agreement encourages signatories to take action to implement and support policy approaches and positive incentives for activities relating to reducing emissions from deforestation and forest degradation and enhancing the role of conservation and forest carbon storage in support of sustainable forest management.

The COP21 agreement also makes it clear that adaptation is urgently required. It calls upon signatories to ensure that education, training, increased public awareness, participation and improved access to information are adequately considered at the national level in capacity-building to strengthen the global response to the climate threat.

As noted above, the agreement also acknowledges that governments can't do all of this alone. The achievement of any meaningful level of climate security will require the coordinated action of all official parties as well as non-party stakeholders, including civil society, the private sector, financial institutions, cities and other subnational jurisdictions, local communities, and Indigenous peoples.

Though still non-binding, Article 7, which calls for formalizing and implementing national adaptation strategies, is very strong. This article makes it clear that adaptation action should follow a country-driven, gender-responsive, participatory and fully transparent approach, taking into consideration vulnerable groups, communities and ecosystems, and should be based on the best available science and, as appropriate, traditional knowledge. In addition, the article calls for sharing of information, good practices and lessons learned and for the further strengthening of scientific knowledge on climate, including research, methodical observation of the climate system and the development of early warning techniques that will inform and support decision making.

Article 7 also establishes the role of the UN's 2030 *Transforming Our World* sustainable development goals in averting, minimizing and addressing loss and damage from extreme weather events.

What happened in Paris should be of particular interest to the private sector. The agreement calls for the enhancement of linkages and creation of synergies among mitigation, finance, technology transfer and coordination of non-market approaches to sustainable development. This agreement is all about opportunity linked to hastening the transition to renewable energy.

The Paris agreement also makes it very clear that climate security cannot be achieved without the cooperative engagement of average citizens in tandem with the full support of the private sector. What is implied but not said is that individuals and corporations with amassed wealth are going to have to put that money to work in service of the planetary good if they are to protect the sources of that wealth or ensure it will have the same meaning in a massively changed world.

But even the most optimistic interpretation of the Paris agreement has to be tempered by a realistic judgment of human nature. Though we are doing an ever better job of characterizing and depicting the damage we are doing to the biodiversity-based planetary life support system upon which our civilization depends, we appear incapable, at the moment at least, of adequately slowing that damage. After decades of tough sledding with respect to advancing climate change mitigation and adaptation, we find ourselves starting all over again from only 100 metres farther down the road toward meaningful action. We can only hope that this time the results will be different.

The first test of the durability of the Paris agreement will be to see if the parties actually ramp up their carbon reduction programs between now and 2020. The second indicator will be to see if the $100-billion climate fund gets topped up by 2020. The final test will be whether we can keep our Paris promises and build on them beyond 2020.

"The climate conference in Paris," said the French

ambassador to Canada, Nicolas Chapuis, three weeks before the conference began, "is an opportunity to put out the fire that is burning our house down." Flames were not seen shooting from the roof at the close of the conference exactly, but clearly the fire is still smouldering inside our house, and hope remains that this fire can be extinguished.

In January 2016 the World Bank published a report about the cost of meeting goals 6.1 and 6.2 of the 2030 *Transforming Our World* agenda, which respectively address targets linked to water and sanitation. The three major findings of the World Bank report were encouraging. The first was that current levels of financing can cover the capital costs of achieving universal basic service for drinking water, sanitation and hygiene by 2030, provided resources are targeted to the needs. The second principal finding was that the necessary capital investments to achieve the water supply, sanitation and hygiene aspects of targets 6.1 and 6.2 amount to about three times current investment levels. The report's third finding observed that sustained universal coverage will require more than just money: financial and institutional strengthening will also be needed to ensure that capital investments translate into effective service delivery.

The Paris agreement is infinitely better than what we were left with after Copenhagen in 2009, but despite fine words and high aspirations, when you separate the real hope from the hype, what is missing is a binding common commitment to act. We are on the river of no return and urgently need clear action now to invest the Paris agreement with the substance it presently lacks. As with all of the UN's *Transforming Our World* sustainable development goals, that substance has to manifest itself at national and subnational levels alike. It is at those levels globally that all hope of addressing the global climate threat presently resides. It is what happens now that will determine humanity's future on this planet.

Investing the Paris agreement with the substance it requires demands a clear picture of where we stand and the risks we face. Much was made during the negotiations that the agreement would be a way of saving the planet. We should be clear on this; our goal should not be to save the planet. The planet will survive on its own terms with or without us, but unless we change our behaviour, we ourselves might not. The negotiations in Paris were not about saving the Earth but about saving ourselves. And we and our flimsy raft of a climate deal are not out of the rapids yet.

CHAPTER 4

Uncharted Waters: Entering the Anthropocene

The idea that humanity has become a force of nature in its own right, capable of disrupting the function of our biodiversity-based planetary life support system, was first put forth by Nobel-Prize-winning atmospheric chemist Paul Crutzen and colleagues in 2000. Since then, evidence has been mounting in support of the claim that we are entering such a completely new era, with three different definitions of the Anthropocene converging toward a consensus that it should be identified as a unique epoch in the history of the Earth. This new era would mark an end to the Holocene, the epoch that began at the end of the last ice age 12,000 years ago and by some 10,000 years ago had stabilized at a global average temperature that remained essentially the same until humanity, because of its numbers and needs, began to measurably alter the global climate.

DEFINING THE ANTHROPOCENE

The 4.5-billion-year history of the Earth has over centuries of analysis been subdivided into a taxonomy of progressively briefer units and subunits of time called eons, eras, epochs and ages based on geological significance. Whether the Anthropocene qualifies as its own unique epoch in this schema will be determined by the International Commission on Stratigraphy, which is not expected to announce its decision until late 2016 or 2017. But there is compelling evidence, based on geological stratigraphy alone, that human activity has expanded globally to such an extent that our impacts now clearly rival natural geological processes in their effect. In other words, effects humans have had on the planet can now be readily observed in rock strata globally.

While there is still debate as to exactly when the Anthropocene actually began, the clearest stratigraphic evidence of a discernible change converges around 1945, widely seen among scientists as being the onset of the "Great Acceleration" of human effects on natural geological processes. That year stands out boldly in the geological record as the time when a new and easily recognizable layer of radioactive materials spread out over the entire surface of the Earth. These materials were introduced into the atmosphere by the advent of nuclear weapons, and they have created a clear marker for geologists for all time that distinguishes this epoch from the preceding one. Some Earth scientists, in fact, believe that current human-induced effects are so much greater and more enduring than what can be inferred from the stratigraphic record that rather than being identified as a minor epoch in the planet's history, the Anthropocene should be recognized as an entire new era, succeeding the Cenozoic, which is held to have begun 66 million years ago and continued to the present.

There are, however, strict criteria for delimiting geological intervals, which involve not just isotopic but lithological and paleontological evidence. It will take time for geologists to examine and integrate the evidence contained in surface depositions and ocean sediments from around the world, and until they complete that painstaking work, the Anthropocene will remain – from a stratigraphic point of view at least – a potential geological epoch only.

A second definition of the Anthropocene has

emerged from Earth system science, a multi-disciplinary research effort that brings together contributions from geochemistry, atmospheric chemistry, global ecology, oceanography, hydrology, glaciology, climatology and other related fields with the goal of achieving a shared systems view of the Earth as a total entity driven by interacting energy and material cycles extending from its core to the upper limits of its atmosphere. While such studies are "geological" in an expanded sense, Earth system science does not derive its evidence for the Anthropocene from sediments or rock strata alone. The evidence is based on the delineation of Earth system boundaries that we appear to be crossing. Researchers have determined that a tipping point has been reached beyond which – as Will Steffen describes it – "the Earth system is operating in a no-analogue state." In other words, humanity has never experienced anything like what is happening to Earth system function now. We are no longer talking about the spread of human impacts over the face of the Earth but a shift in the way our biodiversity-based planetary life support system operates. We are quickly entering a world unrecognizable from the one we have known for the past 200,000 years. As Clive Hamilton, Christophe Bonneuil and François Gemenne note in their book,

The Anthropocene and the Global Environmental Crisis, we have become a telluric force on the planet, a force similar to the great low-frequency currents in the Earth that travel over large areas at and near the surface. The suggestion here is that by its numbers and needs, humanity is changing the nature of the Earth in ways comparable to volcanism, plate tectonics, cyclic fluctuations in solar activity or changes in the planet's orbital eccentricity. Some paleoclimatologists have already gone so far as to estimate that as a result of our greenhouse influence alone we may have already suppressed the Earth's natural glacial cycle for the next 500,000 years. The Anthropocene marks our new place in the history of the world. We are now part of the spectral fingerprint of the planet now and for the future of geological time.

A third proposed definition of the Anthropocene is the product of an expanded view of the impacts humanity has imposed on Earth system function by way of global land use and cover changes, urbanization, resource extraction and the production of waste, the combined effects of which are symbolized by extraordinarily high rates of species extinction. According to those who champion this definition of the Anthropocene, this new epoch marks a distinct change in the relationship between people and the natural world. This definition acknowledges a "step-change" in humanity's impact on and relationship with the natural system out of which we emerged as a species.

LIVING IN THE ANTHROPOCENE

The editors of *The Anthropocene and the Global Environmental Crisis* make yet another interesting case: even if the International Commission on Stratigraphy does not find in favour of the definition of the Anthropocene based on humanity's effect on the Earth as evidenced by rock strata and ocean sediments, the two other proposed definitions will still have considerable force.

Hamilton et al. go on to say that whatever date is chosen as the beginning of "this hijacking of the Earth's trajectory," any acknowledgement at all of the Anthropocene turns our world view upside down. Natural and human history, which have largely been taken to be independent from and incommensurate with one another, must now be considered to be one and the same. A strongly entrenched narrative in our society that has pictured humanity as being somehow above global material and energy cycles, with no need to consider the finiteness of the Earth's resources, has proven to be wrong and dangerous. As a society we have to come back down to Earth. All our institutions – the very manner in which we frame our history, culture, markets and economic and political structures – have to be reformed immediately if we do not want global environmental shifts to extend further and further into the future, entraining generations we cannot yet imagine in a mass extinction of earthly life forms not experienced since the disappearance of the dinosaurs. Our climate is now warmer than has existed on this planet for millions of years. Occurring in tandem are other human-induced impacts such as ocean acidification, synthetic-chemical pollution and alterations of global nitrogen and phosphorus cycles. There is concern in scientific circles that we are clearly near tipping points that could push the Earth and conditions that support life on Earth into a new and unrecognizable state.

Hamilton et al. further note that living in the Anthropocene already entails living in an atmosphere whose composition has been significantly changed by the addition of 575 billion tonnes of carbon since 1870. In the future it will mean living in an impoverished and artificial biosphere that has been irreversibly altered as a result of warming, rising and far more acidic oceans, catastrophic storms and unparalleled levels of new and unequal human suffering and conflict. What will be perhaps most different about this world is that no matter what we do to reduce our current carbon footprint, the Earth system will not rebound to what it was in any time frame meaningful to anyone alive today. As the editors of *The Anthropocene and the Global Environmental Crisis* point out, this fact invites serious new inquiry into contemporary ethics, which they argue are hardly adequate to confront the kinds of questions the Anthropocene challenges humanity to address. Contemporary ethics, they argue, in fact trivialize the threat the Anthropocene poses. They note that, almost unbelievably, there are economists who are right now undertaking cost-benefit analyses to determine where humanity should optimally set the global thermostat. There are also engineers out there who have already concluded that the benefits of geo-engineering clearly outweigh the cost of shattering the world, and that uninhibited technical mastery of the planet is humanity's birthright and destiny. Hamilton et al. remark that it is the unquestioning belief in human reason and confidence in the power of modern technology that makes such people and such thinking so utterly modern and representative of our time, and so very dangerous. They note that one of the striking paradoxes of the Anthropocene is that even though we unwittingly find ourselves a principal force in the transformation of the world, we are clearly unable to govern ourselves effectively in the circumstances we face now, let alone run a world in which change is accelerating toward conditions for which we have no experience. We have entered an age in which the irreversible must be governed. We have begun what will be a permanent state of adaptation where we have to make decisions about a future that cannot be known. Politicians don't know what to do.

It was the creation of the modern global energy system that created our civilization. But because so much has been invested in that system, there are many who understandably do not want to dismantle it, despite the risk that further greenhouse emissions pose to the predictability of the global water cycle and the stability of the climate system. Theirs is a very powerful lobby. Current circumstances are perpetuated by disagreement over the projected costs of potential impacts. Economists appear incapable of producing commensurate estimates of the cost of the damage caused by producing each new tonne of carbon dioxide, making strict regulatory taxes unlikely to produce the desired results within the necessary time frames for action. This situation may be changing, however. In a 2014 report prepared for the National Bureau of Economic Research, called "The Causal Effect of Environmental Catastrophe on Long-Run Economic Growth: Evidence from 6,700 Cyclones," Solomon Hsiang and Amir Jina estimated that "under conservative discounting assumptions the present discounted cost of 'business as usual' climate change is roughly $9.7-trillion larger than previously thought."

No nation or global economic power is prepared to voluntarily endure the self-imposed austerity necessary to decarbonize, especially if competing nations and sectors are not seen to be willing to do the same. In a small but highly charged book entitled *Learning to Die in the Anthropocene*, widely published author Roy Scranton argues that no matter how many people take to the streets in protest or in support of targeted action, the real flows of power in the case of climate action transcend government. The public does not control that power; it only consumes it. We need a new narrative if we are to prosper in the Anthropocene.

CONTESTING THE ANTHROPOCENE

Not everyone accepts the notion of an Anthropocene epoch. While the social scientists are making names for themselves by staking early claims on the fresh ground provided by this concept, there is ongoing debate, particularly in the biological sciences, over how, if we adopt this term, we should embrace this new concept. Some highly respected scientists, such as the renowned sociobiologist Edward O. Wilson, argue that those who think we can just walk into the engine room of the planet and take over Earth system function in our own interest don't have any idea of what they are talking about. Wilson warmed up on this subject in a number of earlier books and then took on the entire issue of the false pretences in Anthropocene ideology in a 2016 book called *Half-Earth: Our Planet's Fight for Life*. Wilson doesn't hold back. He makes it very clear that his view is that by caving in to the Anthropocene and its acceptance of human mastery of the Earth system we are engaging in an endgame that will result in the creation of circumstances on this planet that will be unfavourable to any life except microbes, jellyfish and fungi.

Wilson reminds us that the biosphere does not belong to us; we belong to it. He maintains that if we want to remain within the safety margin of Earth system function, we have to preserve half of the Earth's critical ecosystems and the biodiversity they represent or we will not be able to ensure planetary conditions that will be livable, at least in terms of how we define that word today. Wilson doesn't think we have entered the Anthropocene yet, and he notes that even if we want to call it that, the Anthropocene is an inaccurate description of the world as it will exist in such an epoch. Instead, he would name such an epoch the Eremocene, an age of absence of other species, which he imagines as being characterized by the sole presence on Earth of people and our domesticated plants and animals expanding in every direction as far as the eye can see.

More descriptive cautionary names have been proposed, too. Michael Soule suggested the Catastrophozoic Era. Other contenders include "Homogenocene, the Age of Homogeneity," and "Mixocene, the Age of Slime." Because of the prevalence of plastics in our world, ethicist Kathleen Dean Moore has proposed to call this new epoch the Plasticene. In her new book, *Great Tide Rising*, Moore notes that culpable wrongdoing is defined as the knowing and intentional doing of unjustifiable harm. It is therefore unjust, she argues, that the costs of burning fossil fuels in our generation will be borne by the blameless and the voiceless. What from a moral point of view, she asks, could future generations do to deserve what we are leaving them? We have to align with evolutionary processes rather than trying to bring them to an end. The distinction between past processes of extinction and what is happening now is that earlier extinctions were natural. What we are doing should not be categorized as the sixth extinction, because the earlier extinctions were beyond human control or culpability. Moore is of the view that extinctions 1 to 5 should inspire awe. Extinction 6, however, which we are bringing upon ourselves now, should, in her view, call us to moral outrage. The extinction for which we are responsible is murder. We are doing this knowingly, willfully and recklessly. Unlike the dinosaurs, we have a choice.

For all our supposed intelligence, we are the most destructive species in the history of life on Earth. E.O. Wilson addresses the question of why – if extinction is the normal course of species – we should be concerned about the growing rate of extinctions. His answer is that in past extinctions, species didn't usually die at all. They instead evolved into two or more daughter species. The accelerated rates of extinction brought about by human impacts do not allow new species the time required to emerge from the old. We are on the way to being alone on Earth, hence Wilson's view that we are entering not an Anthropocene, an Age of Us, but an Eremocene, an Age of Loneliness.

Wilson reminds us that we know very little about the life forms with which we share our planet. We have identified 20 per cent or less of the biodiversity on this planet at the species level. With species disappearing at current rates, most will vanish before they are even known. We will feel the effect of this loss on biodiversity-based Earth system function without even knowing what caused it. We may be able to identify and catalogue the rest of life on Earth by about 2300, but by then we don't know how much will still exist. Freshwater species diversity is in particular trouble. Some 57 species of freshwater fish are already extinct in North America. The rate of extinction is 877 times the prevailing rate that existed before humanity arrived on the scene in large numbers. For all their passion, human conservation efforts globally have only slowed extinctions by perhaps 20 per cent. The most vulnerable habitats of all, with the highest extinction rates per unit area, include rivers, streams and lakes in both tropical and temperate regions of the world.

Wilson believes that the focus of the contemporary conservation movement on the new Anthropocene ideology is utterly misplaced. The proponents of this ideology argue that traditional efforts to save the Earth's biodiversity have failed. They claim that pristine nature no longer exists and that true wilderness is only a figment of our imagination. Extremists among those of the Anthropocene persuasion go so far as to argue that what is left of nature should be treated as a commodity to justify saving it. They argue that the surviving biodiversity is better judged by its service to humanity than by its contribution to biodiversity-based life-support system function. They advocate the notion that it was, in fact, the Earth's destiny to be fully occupied by humans. The bottom line – and there is always a bottom line in the Anthropocene – is that even wildlife should have to earn its livelihood just like everyone else. This would mean that national parks and other reserves should be managed to ensure they meet the needs of people. But it is not the needs of all people that national parks and nature reserves should serve, only the needs of those of us who are alive today and in the near future. Surviving wild species of plants and animals are now expected to live in a new relationship with humans. Whereas in the past people entered natural ecosystems as visitors, in the Anthropocene envisioned by some, species in remaining fragments of ecosystems will be expected not just to tolerate us but to live among us.

If you have ever wondered why there is tension between the physical sciences and the social sciences and humanities, Wilson gives some cogent reasons. He maintains that various leading proponents of the Anthropocene are as free of facts as they are of fear. He cites Erle Ellis, for example, an environmental scientist at the University of Maryland who is touted as something of a prophet of this new world order. Wilson notes that remarks such as Ellis has made essentially illustrate what is wrong with the credo of the Anthropocene extremists, essentially saying "stop trying to save the planet; nature is gone; you are living on a used planet; if this bothers you, get over it."

Wilson obliterates such suppositions. He in essence accuses the Anthropocene extremists of biotic cleansing: making species killable and legitimizing extinction through the commodification of nature. He begins his counter-argument by challenging the notion that wilderness no longer exists. The extremist Anthropocene view is that since wilderness no longer exists except in our minds, we should be comfortable exploiting it in our own interests. Wilson reminds us that wilderness means a large area in which natural processes unfold in the absence of human intervention – essentially those places where life remains self-willed and self-regulating. He goes on to name and describe dozens of the great wildernesses that still remain on Earth. He points out how dangerous it is to allow Anthropocene extremists to redefine the natural world out of existence when it is obvious that the people who are proposing such ideology have clearly never experienced wilderness. Wilson's point is that if you have spent your life in a city, it might be easy to imagine an entire world given over to humanity. But if you haven't lived your whole life in a city and have actually experienced wilderness, then your worldview will be quite different. On this matter, Wilson quotes the great explorer–naturalist Alexander von Humboldt (1769–1859): "The most dangerous worldview is the world view of those who have not viewed the world."

Wilson concedes that most Anthropocene world views are not malicious but instead a product of well-intentioned ignorance. He maintains there is simply so much we don't know. Among these matters is the not insubstantial issue of invasive species. Wilson utterly rejects the idea put forward by various Anthropocene populists of turning the entire world into what Emma Marris described as "a global, half-wild rambunctious garden tended by us." Aside from the fact that we are the most invasive of all species, invasive species are a much bigger problem than the Anthropocene extremists appreciate. A small percentage of the world's invasive species, Wilson points out, causes a disproportionate amount of damage. These species include the Asian termite (which ate New Orleans), the gypsy moth, the emerald elm beetle, the Asian carp and zebra mussels. We need a much deeper understanding of ourselves and the rest of life, Wilson maintains, before we give in to the philosophical absurdities being put forward as the underlying pretext for the Anthropocene.

Wilson points out that the promise of alien species somehow filling in gaps and balancing the extinction of local native species through some miraculous act of hybridization is ridiculous. He notes that despite all the recent hype suggesting otherwise, invasive species will *not* settle down with native species into stable "new ecosystems," as some contemporary writers have supposed. He reminds us that the natural restoration of biodiversity will require evolution over five million years or longer, which he observes is more than an order of magnitude longer than it took our own species to evolve. In the meantime, society will have to bear the cost of the damage caused by invasive species, which will not be insubstantial. That cost in the United States alone, Wilson claims, is $137-billion a year.

To save biodiversity you have to obey the precautionary principle, which Wilson says Anthropocene extremists utterly ignore. As articulated in *Wikipedia*, the precautionary principle (or precautionary approach) to risk management states that if an action or policy has a suspected risk of causing harm to the public or the environment, then, in the absence of scientific consensus (that the action or policy is not harmful), the burden of proving non-harmfulness falls on those taking an action that may or may not be a risk. The principle is used by policy makers to justify discretionary decisions in situations where there is the possibility of harm from making a given decision (e.g., taking a particular course of action) when extensive scientific knowledge on the matter is lacking. The principle implies that when scientific investigation has found a plausible risk, there is a social responsibility to protect the public from exposure to harm. These protections can be relaxed only if further scientific findings emerge that provide sound evidence that no harm will result. Much Anthropocene ideology, Wilson notes, is based on computer models. But how, Wilson asks, can such models bear any resemblance whatsoever to reality when we don't even know the identity of, let alone the ecological role played by, even a fraction of the species that turn the finely tuned engines of global energy and materials cycles? The Anthropocene extremists have no idea what they are talking about. It is projected that two-thirds of the species on Earth will remain unnamed and unknown. How can we be masters of the Anthropocene when we don't even know what we are masters of?

In Wilson's estimation, writers and spokespersons for the Anthropocene are as innocent of the meaning of biodiversity at the species level as the 19th century phrenologists were of neurobiology when studying the shape of the human skull. In accepting ongoing loss of species and asserting that wilderness doesn't exist, we are permanently dooming ourselves to ignorance of the true meaning and function of biodiversity. In accepting such ignorance as a precondition for our heroic self-centred entrance onto the Anthropocene stage, we are also accepting risks we barely understand. The fact is that we have no idea what warmer mean global temperatures will awaken. Wilson notes that numerous kinds of bacteria – many of which we have yet to identify – obey a scientific principle first proposed in 1934 by the pioneering ecologist Lourens Baas Becking (1895–1963). "Everything is everywhere," Baas Becking pronounced, "but the environment selects." What he meant was that a large fraction of genetic forms occur around the globe, but most of these lie dormant. Together, however, these forms in essence are a seed bank within which each species begins to multiply when the environment changes in ways that suit its DNA preferences. These sleeping cells will awaken when presented with the right acidity, the right nutrients and the right temperatures.

What Wilson is telling us here is that a vast storehouse of ancient bacterial DNA exists, ready to leap into new bacterial forms with the slightest change in Earth system conditions. We have no idea whatsoever what new bacterial and viral agents might come into instant existence as a result of the warming and changing of Earth system conditions we are bringing about. These latently active species are all around us, too small even to be seen in most microscopes. If we decide to blunder into the Anthropocene, we will be seeing life forms that we have never experienced before, many of which we will fear. Ebola, HINI and the Zika virus are just a hint of what we may awaken in the future as we change the conditions of life on Earth. This in itself, Wilson argues, should be reason enough for humanity to ignore the entreaties of Anthropocene extremists and abandon our hubris. Our goal instead should be to back quietly and humbly but actively out of the dominant role we are playing without any qualifications whatsoever on the planetary stage.

The danger here resides in trustingly and without due consideration accepting that we have crossed a line and entered into a human-dominated geological epoch without examining whether or not we should reject the role we appear to be playing in Earth system decline. Wilson argues that we still have time to take a more humble and less dangerous course that does not presume human dominion over a world we barely understand. He challenges the authority Anthropocene extremists have assumed without permission to claim that since in their view the loss of biodiversity in the limited manner in which they understand it is beyond the point of no return, it is reasonable to argue that protecting such biodiversity doesn't matter anymore.

How does an unknowing and sometimes gullible

public keep from sliding down the progression of intellectually sloppy steps that mark the moral slippery slope Anthropocene extremists want humanity to take in claiming that even without knowing the full extent of the Earth's biodiversity, it would be perfectly reasonable for us to take over full control of the remaining fragments of the diminished natural world? Wilson argues we should be challenging the shallow premises of the nature-is-dead defeatists and the ill-informed and short-sighted ideologies of the Anthropocentrists. He maintains that even though extinction rates are soaring, a great deal of the Earth's biodiversity can still be saved, and he explains where and how that can happen.

Wilson concludes by making the point that in the concept of the Anthropocene, all we are being offered is that we will be saved somehow by a variety of as yet undisclosed acts of human ingenuity. In his view, entering the Anthropocene on these terms is little more than drifting mindlessly toward wishful ends. He points out there is more at stake here than we know, and that it is important to understand that in taking the Anthropocene gamble there is no going back. Such a decision would be irreversible. In giving up on protecting the biodiversity-based Earth system planetary life support function, there will be no going home again. We will be responsible for replacing elements of the Earth system function we have lost. Are we ready for that? Do we want that? E.O. Wilson urges us not to accept the notion that pauperized ecosystems and diminished Earth system function cannot be restored. Instead, he offers what he thinks should be humanity's next new metanarrative. It is easy and simple: do no further harm to the biosphere.

CHAPTER 5

Bringing It Home: The Potential Impact of Climate Disruption on Our Way of Life in North America

Those who would deny humanity's effect on the composition of the Earth's atmosphere make much of the fact that climate has changed often before and during the course of human history. We have clear evidence that this has been the case in the Middle East, North Africa, China, India, the American Southwest and more recently at high latitudes in Europe, Greenland and North America. But each of these disruptions occurred when human populations were relatively fragmented and orders of magnitude smaller than they are today. We are threatened now by what could potentially be abrupt climate change on a global scale at a time when there are more people on Earth than ever before. The world is largely inhabited – crowded in many places – and there is nowhere left that is not under one form or another of vigorous jurisdictional territoriality. And there will be fewer such places left as sea level rises. As we recently saw in Europe in the wake of a five-year drought that helped spark the civil war in Syria, we have to prepare for human migrations on a scale not witnessed since the end of the last world war.

If there are lessons we might learn from what is happening in the rest of the world, our attention would likely be directed to the brutal fact that unless we act very quickly, we – all of us – are going to have to plan for what it is going to be like to live in a world that on average may be as much as 4°C warmer. In other words, failure to act in effective and timely ways now will mean that over time we will be faced with making our way and finding happiness and security in a world that will resemble less and less the one in which we evolved as a species.

There is nowhere in North America – except perhaps the Arctic – that is likely to change more than the centre of the continent. At the moment, the most obvious effect of changing hydro-climatic regimes on the central prairies of North America is the poleward advance of subtropical storm tracks. But there is another problem: the spectre of deep and persistent drought. We don't know if there are other invisible thresholds that will be crossed as mean temperatures continue to rise. Temperatures on the prairies are expected to warm by between 5° and 8°C. We don't know what threshold of temperature increase will cause the hydro-meteorological coin to land permanently with its dry side up. All we know is that if temperatures continue to rise, sooner or later it will.

Even small mean temperature increases result in far more days of extreme heat. In the pre-industrial world you could expect one extreme heat day for every thousand days. With warming today of only 0.85°C, the prospect of extreme heat days is four to five times higher than in pre-industrial times. But what happens when the temperature continues to rise? The number of extremely warm days rises exponentially, not linearly. With 2°C of warming, the number of extreme heat days is multiplied by more than five times over current conditions. At 3° of additional warming, the prospect is that the number of extreme heat days – days with extreme temperatures beyond anything we have experienced in our time – grows by an order of magnitude. We have no idea what will happen if the 5° to 8°C of warming projected for the Central Plains actually materializes. Clearly, there will be profound effects on agriculture. Some of these changes will be beneficial, at least at the outset, others less so. It is the net effect of warming on agriculture on the Central Plains of North America, however, that is of greatest concern.

A NEW GREEN REVOLUTION
FOR A DRYING PRAIRIE

One of the keynote speakers at the Canadian Climate Forum held in Ottawa in 2015 was Dr. David Sauchyn, who holds the research chair in water resources and climate at the University of Regina. Dr. Sauchyn spoke on the topic of the availability of agricultural water in a warming climate. He began by comparing conversations in Chile and Canada regarding water and food security, which he demonstrated have common uncertainties we all need to address. He then showed photos of the Canadian prairies, pointing out that this is a landscape of adaptation – successful adaptation, at least to date. He went on to show research findings that indicate an increase in the number of growing days and heat days, together with greater water availability in winter but less in summer when it is needed most. He then pointed to research outcomes that suggested that a potential 228 per cent increase in crop yield could be possible in some parts of the

Canadian prairies, based on these trends. But he noted that annual averages were not the whole story. The range between drier and wetter conditions is growing, which means when dry periods occur they are likely to be more extreme. Sauchyn then showed the mean average water flows on the Canadian prairies over the past 1,100 years, which made it clear that droughts of a century or longer have occurred and will occur again, but in a warmer world.

Dr. Sauchyn went on to outline his five-year-long conversation with farmers and government agricultural experts from five countries. Farmers and ranchers, Sauchyn noted, recognized the role of adaptive planning, awareness and education and the need to review and refresh plans. They recognized the need to be proactive but also pointed out that a technical gap exists that can only be filled if research into agriculture were restored so that the scientific needs of farmers could be met. Sauchyn recommended a single coordinating agency that could meet all the research-related needs of agriculture. Dr. Sauchyn urged the government of Canada to create or recreate such an agency in the image and spirit of the Prairie Farm Rehabilitation Administration, the function of which was much reduced under an earlier government.

In both the American Southwest and the Central Great Plains the risk of a 35-year drought occurring by the end of the present century has been calculated to be as high as 80 per cent if climate effects go unmitigated. What we are seeing here and globally is that if we want flood and drought resilience in our cities – or anywhere else – the people in cities are going to need agriculture's help. If water, food and climate are to remain in our grasp, agriculture must become restorative as well as productive.

What we may need is another agricultural revolution, one in which society agrees to pay farmers not just for crops but for perpetuating critical Earth system functions over the ever-expanding lands now under cultivation globally. We know what direction we should head, and that is toward improved soil health. Healthy soils remain humanity's first and foremost water purification system and a powerful aid in flood and drought resilience. But soil does something else besides grow forests, supply food and absorb and purify water. It stores carbon. It is now estimated we have already lost as much as 80 billion tonnes of carbon from our soils through inappropriate agricultural practices and short-sighted land use. Keeping carbon in the soil, therefore, may be one of humanity's most important priorities.

But even more broadly than carbon storage alone, this second green revolution should focus on integrating the entire nexus of water, food and climate security. What's more, North America should be a leader in this, for such a revolution is no longer just optional when it comes to farming the Central Plains. Current industrial agricultural practices are not just unsustainable but self-terminating in terms of their effects on water resources. Rising temperatures will make the way we currently grow food on the prairies in a warming world a trade-off many North Americans are unlikely to accept. If there is one single example that symbolizes the kinds of unintended consequences warmer temperatures are going to bring in their wake with respect to agriculture, it is what is presently happening to the largest lake on North America's Central Plains.

LAKE WINNIPEG: TOO MUCH OF THE WRONG KIND OF GREEN

It appears we keep crossing threshold after invisible threshold with respect to changes in Earthly hydrological, biological and climatic interactions. And even though we know this will result in step-like changes in the conditions we rely on for the hydro-ecological-climatic stability that is the very foundation of our social, economic and political stability, we appear to be able to identify these thresholds only after we have already crossed them. The latest one we have discovered is symbolized by the changing condition of Lake Winnipeg.

At 25,514 square kilometres, Lake Winnipeg is the sixth-largest lake in Canada. Its watershed is a huge basin that straddles four Canadian provinces and two US states. While not a transboundary water body in its own right, the lake is fed by the Red River, its major source, which flows northward out of the United States.

The urgency of meaningful action to address the deteriorating ecological condition of Lake Winnipeg, and the threat it poses to the economic health of surrounding communities, is accelerating. The scale of the problem appears to be growing both in extent and complexity. Algal blooms as large as 17,000 square kilometres are now appearing in the lake.

Renowned scientists such as Dr. David Schindler at University of Alberta and Dr. John Pomeroy at University of Saskatchewan have demonstrated that the basin has, in fact, crossed an invisible hydro-climatic threshold into a new hydro-meteorological regime. While the western part of the basin has gotten drier and is delivering less water and nutrients, the eastern portion has done the reverse. In this eastern region of the basin it appears we can expect more frequent large-scale flooding that will mobilize more nutrients, not just into Lake Winnipeg but into all of Manitoba's like water bodies. Members of Manitoba's Lake Friendly Alliance have been tracking what has happened to Lake Winnipeg in the barely two years since the alliance was formed to address eutrophication and other threats to lakes, streams and rivers in Manitoba.

Manitoba is not alone in facing the problems associated with algal blooms. Independent research identified the presence of harmful cyanotoxins such as those that have appeared in Lake Winnipeg's blooms in 246 other lakes across Canada. While recognized as a regional problem for the past 50 years, eutrophication is clearly becoming an issue nationally.

As if the algae weren't enough, zebra mussels were later discovered in both the Red River and Lake Winnipeg, a development that will complicate every effort to restore the health and maintain the biodiversity of the big lake and similar water bodies throughout southern Manitoba. The issue of invasive species cannot be ignored. As pointed out earlier, the damage such species cause in the United States alone has been calculated at $137-billion a year.

Neither did it help the reputation of the province much when an international fisheries watchdog agency later proclaimed to the world that because of what it perceived as poor fisheries management, consumers should not buy fish from Manitoba's largest lakes.

The urgency of dealing more quickly with the Lake Winnipeg situation was further brought home in early December 2015 when the newsletter of the Lake of the Woods Water Sustainability Foundation published satellite images of the unprecedented extent and long duration of algal blooms in that nearby lake system. The presence of such large blooms in Lake of the Woods, in addition to reports of the return of eutrophication in Lake Erie and other Great Lakes after decades of successful remediation, makes it clear that the problem may be more widespread than we thought. What happened next confirms this.

Hard on the heels of the Lake of the Woods report, a global survey of hundreds of the Earth's lakes, published in *Science*, revealed that climate change is causing lakes to warm faster than the oceans or the air above them. One reason is that warmer winter temperatures are producing less ice on lakes that normally freeze over. Reduced ice coverage in turn increases the amount of sunlight lakes absorb. These rising temperatures will not only exacerbate problems associated with eutrophication, but may also speed the conversion of carbon-rich organic matter in lake sediments into methane and carbon dioxide, gases that, once released into the atmosphere, exacerbate global warming.

What's more, further research has demonstrated that lake temperatures in Canada appear to be warming at twice the global average. We can all see where this is heading. No lake in Canada will be safe much longer from the combined effects of impacts we have brought about by changing the composition of the Earth's atmosphere while at the same time dramatically altering land-use and agricultural practices in Canada and globally.

Less than two months after it was reported that lake temperatures worldwide were rising, an even more alarming report offered evidence that algal blooms are also expanding in the world's oceans. In February 2016, researchers in Alaska with the US National Oceanic and Atmospheric Administration announced that more than 900 samples from Arctic marine mammals that were either harvested or died while stranded revealed algal toxins in predator species including bowhead whales, fur seals and sea otters. These neural poisons – domoic acid and saxitoxin – are deadly in high doses and were found in 13 marine mammal species in habitats from southeastern Alaska to as far north as the Beaufort and Chukchi seas. Researchers were surprised by the findings. According to one of the lead NOAA fisheries scientists, Kathi Lefebvre, quoted in a 2016 *Washington Post* story by Ryan Schuessler, "We did not expect these toxins to be present in the food web in high enough levels to be detected in these predators." But the discovery goes a long way toward explaining strange marine life die-offs in the Arctic, including the death of more than 30 whales in Alaska in 2015 and thousands of dead birds that began washing ashore in Prince William Sound in January 2016. Arctic waters are warming; sea ice is melting. The greater exposure to sunlight accelerates algae growth, and with algal blooms come deadly cyanotoxins. The researchers noted their study added further evidence that warming ocean temperatures are allowing algal blooms to extend into Arctic ecosystems, with significant consequences for communities that rely on the sea for food and livelihood.

What is presently being missed here, however,

are the larger implications. Rather than control our own swelling numbers, humanity instead mounted a green revolution. With the goal of feeding our growing billions, we overfertilized the world – but wastefully.

FERTILIZER: PHOSPHORUS FINDS A WAY

Phosphorus is naturally present in soil in a chemical form known as phosphate. But because phosphate is found in relatively small amounts in most soils, it often acts as a limiting factor in terms of crop production in grasslands and other agricultural systems. Adding phosphorus-containing fertilizer to the soil can boost a system's productivity – but that kind of fertilizer is becoming less and less readily available. That's because such fertilizers require large amounts of phosphate rock, which must be mined. And there's a limit to how much of it is left on the planet. Some researchers estimate that we waste about 80 per cent of the phosphate we use specifically for food production, and that these losses occur at all stages, from where the stuff is mined to where it is applied to the fields in producing food.

Now we are finding out what happens to all that wasted phosphorus fertilizer in a warming world. Lake Winnipeg is just the tip of the rapidly melting iceberg. We wanted a green revolution, and now we have one. But it is neither the one we expected nor the one we desired. Yes, we are managing to feed the starving billions, but we are feeding something else, too. As the character Dr. Ian Malcolm famously said in the film *Jurassic Park*, life finds a way. It appears that in the world's warming waters, lower life forms are conspiring to take their planet back. The great irony is that for all the grief they cause today, cyanobacteria achieved such numbers early in the Earth's history that they produced the "great oxygen revolution," during which the oxygen released as a result of their metabolism became so abundant it changed the composition of our planet's atmosphere. As oxygen was poisonous to many of the other early forms of bacteria, they were driven to the depths of the ocean, to the extreme oxygen-free environments of undersea volcanic vents or to the anaerobic plumbing of hot springs. It was in the oxygen-rich atmosphere created by the cyanobacteria that all modern life forms evolved – including our own.

Cyanobacteria continue to produce a significant proportion of the oxygen we breathe today. But suddenly they have gone rogue. The Anthropocene be damned, with humanity as an unwitting ally, the algae of the world appear poised to poison the planet back to an earlier geological era. Who could have expected we were so close to such a threshold? But we can learn from this. What we are witnessing now is what warming does to water and how life responds. Whether in rivers and lakes, in the global ocean or circulating in the atmosphere, pay attention to the warming of the world's waters, for it will be through water that climate change will express itself most fully.

The question now becomes this: If we know all this is happening, why don't we do what is necessary to address problems of this magnitude? This area of public discourse is now getting more attention than ever, not just from social scientists but from political commentators and journalists.

CHAPTER 6

Human Nature and Organizational Inclination:
Habits That Are Holding Us Back

THE POLITICS OF FEAR

In his book *Learning to Die in the Anthropocene*, Roy Scranton maintains that every thinking person knows what the problem is, but it is too big to address: it is us. He also points out that global warming presents us with no apprehensible foe. The enemy is not "out there." It is in us, and not as individuals but as a collective. Scranton calls it a system, a hive. Though we may fully see and understand the climate threat, nobody has the tools, the clout or the conceptual framework to solve the problem. All we can do, according to Scranton, is fumble around with political machinery that doesn't work.

Politicians, Scranton observes, are trying to head off this strife by making us servile and weak by constantly making us fearful of one another and the future. People, he notes, will do almost anything to avoid being afraid. When we are afraid we discharge our fears by passing them on, "retweeting the story, reposting the video, hoping that others will validate our reaction, thus assuaging our fear by assuring ourselves that collective action has been alerted to the threat." With every retweet, Scranton argues, we become stronger resonators and weaker thinkers.

Other times, Scranton argues, we simply react with aversion. Often our aversion takes the form of more consumption. We seek out positive reinforcements, pleasurable images and videos, something we find funny – anything to ease the fear. By providing "fierce economic competition, hyper-violent television, professional sports mixed with occasional protests to let off steam," what governments and corporations do, Scranton says, is create a cruel optimism. He believes, however, that once the social fabric begins to tear, we risk unleashing not only rioting, rebellion and civil war but "homicidal politics the likes of which should make our blood run cold."

To turn our situation around, Scranton believes we have to become far more thoughtful as individuals. Our individual sovereignty will depend on letting our "integrated impulsion" die in us. We have to recognize that we are free only to the extent that we are willing "to interrupt escalations of group-think and immunize ourselves against epidemic infections of popular opinion. We have to resist and redirect ways of excitation that flow through our society."

Scranton argues that as we struggle, socially awash in fear and aggression, to face the catastrophic self-destruction of global civilization, we must more than ever practise critical thought, contemplation and philosophical debate and keep posing impertinent questions. He offers that we must continue always to inculcate thoughtfulness in those around us by teaching slowness, attention to detail, careful reading and meditative reflection: "We must suspend our attachment to the continual press of the present by keeping alive the past, cultivating the info-garden of the archive, reading, interpreting, sorting, nurturing, and, most important, reworking our stock of remembrance." What might save us in the Anthropocene, Scranton observes, is what saved us when our existence was threatened in the Holocene: memory. Our meaning-making and heritage-sharing, however, require constant reworking. We need to build arks, Scranton says, "not just biological arks, to carry forward endangered genetic data, but also cultural arks, to carry forward endangered wisdom." The fate of the humanities, as we confront the end of modern civilization, may in the end be the fate of humanity itself.

WHAT IS TO BE DONE, THEN?

Renowned philosopher and ethicist Dale Jamieson is of a different view. His book, *Reason in a Dark Time*, written over a period of several decades and published at last in 2014, is widely considered to be one of the most important books on the climate issue in the last several years. Though it came out before the Paris conference, what it contains is every bit as much, if not even more, relevant in the wake of those negotiations.

The book is valuable in that it provides an excellent summary of the history of the climate problem. It is very useful also in its succinct and highly articulate analysis of the scientific, social, economic and political obstacles to action. Jamieson goes to great pains to clearly define the limits of contemporary economics and economic theory in addressing the climate threat. He also talks about the limits of ethics and contemporary morality in adequately contextualizing issues of harm and responsibility when it comes to hydro-climatic change. Jamieson insists that in light of our failed capacity to address the problem, we are now faced with planning for how we are going to live in the Anthropocene.

Jamieson is very careful also to put us on notice regarding the kinds of dangerous directions we may be tricked or forced into taking as climate change accelerates. He explains, for example, the dangers of allowing proponents of geo-engineering to have their way. He outlines what the impacts on our global economic system and already fragile international relations might be if we accede to radical, untested means of dealing as cheaply and quickly as possible with a growing number of climate-related emergencies. This is not a book for the ethically faint of heart, however. The problems we face are a lot deeper than what we see on the surface of North American society. After decades of careful historical and philosophical analysis, Jamieson concludes that the America of today is not just cynical but may well be heading in the direction of breakdown with respect to Enlightenment intellectual values. He believes, with a good bit of evidence to back his view, that in the United States at least, if not more widely, science is a threat to the established belief system. Rather than proving the truth and utility of ideas, as was the Enlightenment goal, we now market and brand ideas and manipulate consumers into buying them. In so doing we have condemned future generations to living in a warming world for at least a thousand years. While recognizing that climate change may undermine our sense of an ordered and just world, Jamieson maintains that it is still possible to find meaning in the Anthropocene. While there is no question that climate change threatens a great deal of what is important and enduring about our society no matter where and how we live, in Jamieson's estimation, "it does not touch what ultimately makes our lives worth living: the activities we engage in that are in accordance with our values." Jamieson maintains that while contemporary green virtues may not be an algorithm for addressing the challenges of the Anthropocene, they can provide guidance for living gracefully in a changed world. By living according to appropriate values, Jamieson believes, reasonable people will still have meaningful lives in the Anthropocene.

Jamieson does, however, point very practically in the direction toward which action must follow. Long before the Paris climate summit, he offered seven priorities for a society committed to responding to climate change. The first priority nationally must be to integrate adaptation with development. The second priority is to manage water better, with the goal of protecting, restoring and increasing terrestrial carbon sinks while still honouring the broadest range of human and environmental values. The third is to develop full-cost energy accounting that takes into account the entire life cycle of producing and consuming any given unit of energy. The fourth is to raise the price of emitting greenhouse gases to a level that roughly reflects their costs. The fifth priority is to force the adoption and diffusion of existing and emerging technologies. The sixth is to substantially increase research in renewable energy and carbon sequestration, especially as it relates to improving agricultural soil health and prudent forest management. The seventh priority is to plan for the Anthropocene, the brave new epoch we appear to be entering in which we are going to have to take direct and personal responsibility for managing and restoring vital elements of the Earth system.

Our most immediate action, Jamieson recommends, might be to stop burning coal and plan for the end of coal mining. This is where what happened in Paris really starts to hit home. How you might suddenly stop burning coal any time soon in a state or province that relies heavily on coal-fired generation of electricity challenges the imagination.

Whether we like it or not, or even believe it,

Paris also put the writing on the wall with respect to the future of oil and gas. While nothing is going to happen overnight, plummeting oil prices at the time of this writing and the disastrous effect they are having currently on Canada's national economy, for example, foreshadow the kinds of difficulties the restructuring of the global economy around renewable energy will create for some in the future. For many of us – rich and poor alike – it will take a great deal of imagination and ingenuity to even come close to meeting the Paris objectives without significant effects on our way of life.

Others, including many social scientists, are not confident that more information and greater individual thoughtfulness are going to be adequate to penetrate the thick skin of the status quo when it comes to addressing the climate threat. Facts, they maintain, do not speak for themselves. They argue that a public mindset is just that: a set mind. These voices assert that stability is at the core of human identity, and that in order to create alternative sensibilities, something has to be erased from the existing mindset for a new one to take hold. What is required, these scientists claim, is nothing less than a full "cognitive collapse," which can only come about as a result of coming into close proximity to catastrophe. Social scientist Gudmund Hernes is one of the leading proponents of such views.

EVENT-DRIVEN CHANGE

Gudmund Hernes is one of Europe's leading behavioural scientists and an expert on human responses to complicated problems like global warming and climate change. In a high-level report to the prime ministers of Europe's Nordic countries, Hernes demonstrated that we do not change our world view just because it is logical to do so. As he notes in his book, *Hot Topic – Cold Comfort*, changes in world view in the past have been brought about not by incremental reason but by changes in the world itself – in this case, changes we have brought about ourselves. Public mindset change, Hernes observes, is driven by events. It is also important to note, however, that these changes do not occur in a linear way, with concern about recurring disasters leading suddenly and directly to new perspectives.

Hernes cites Arthur Koestler, who in 1958 observed that science, which often informs the advancement of public perceptions, does not progress in a straight line either. Rather, it takes a zigzag course through history, which without exaggeration is influenced so much by "collective obsessions and controlled schizophrenias" that the course of human thought resembles more the behaviour of a sleepwalker than the reasoned logic of a computer. The course of human understanding related to the climate threat has been further complicated and set back by the active agitation of so-called climate realists and true climate deniers who are paid to confuse the public about the scientific consensus and downplay risk associated with changes we are bringing about in the composition of the Earth's atmosphere.

"CLIMATEGATE"

The effect of orchestrated opposition to the rising public profile of climate change issues was clearly demonstrated just before the climate conference in Copenhagen in 2009 by the staged release of thousands of emails from scientists at the University of East Anglia in an effort to assert that the researchers had withheld information and suppressed contrary views. This carefully orchestrated propaganda effort succeeded in undermining public confidence in the science (and the scientists) at the heart of climate research at a critical time in the negotiation of an international agreement on climate change and delayed the global policy response by years.

Following "climategate," as the alleged fraud

came to be called, there was a rise in the use of expressions like "climate fatigue," a term which suggested that the public was becoming fed up and indifferent to the prophesies of these climate Cassandras and to the dire warnings of all the environmentalists who for years had been relentlessly crying wolf. The failure of the Copenhagen conference to produce any meaningful action was one measure of the success of the climategate subterfuge. It took the scientific community years to prove the charges false, but in the meantime the idea that the public could simply get tired of the entire climate issue and thereby dismiss it became mainstream.

While there may have been increasing recognition that the global environment was being harmed as a consequence of human numbers and activities, it became widely accepted, in North America at least, that inadequate knowledge, lack of scientific consensus, the absence of shared urgency and the failure to agree on an ethical foundation upon which to act were reasons enough to ignore the problem. But the problem did not go away. Events continued to demonstrate that our new global climate situation could not be ignored. Just as Gudmund Hernes predicted, climate-related disasters continued to jolt the public mind, undermining preconceived notions of societal stability in ways that demonstrated dangerous inconsistencies in our faltering world view. Events are showing that if you are tired of climate issues now, you haven't seen anything yet. The clear mismatch between the range of impacts that are being caused by human activities and the capacity of the organizations we have put in place to cope with them is now glaringly obvious. Events have begun to dissolve and uncouple public confidence in our basic institutions at such a deep level that many now openly argue that, in entering the Anthropocene, we have no choice but to crystallize a new world view into a new societal gestalt. If we don't, we could collapse as a society.

THE PSYCHOLOGY OF DENIAL

Gudmund Hernes clearly recognizes the mechanisms by which people stick to erroneous or unsupportable misconceptions, even in the presence of strong evidence to the contrary. He understands how selective perceptions work and how selective memory and wishful thinking lead to comfortable beliefs. He knows that the human capacity for rationalization can make us into cognitive escape artists. But Hernes also knows that we react to disaster by reacting to one another. We need to recognize that what were once slow-moving and linear climate-related events are rapidly metastasizing into a chaotic future we neither planned nor desire. We also know now that we are part of a system that is not likely to revert to equilibrium when disturbed, a system that when jolted may very well spiral out of control. How many and how widespread will disasters have to become before the social transformation required to prevent societal collapse takes place? We can no longer hesitate to react. What the world awaits now is a positive, new and very different narrative encapsulated in a potent new metaphor that will penetrate all our opinion networks and show the way to a survivable future. It should be the goal of every thinking person to help crystallize that narrative and articulate and share that urgently needed new metaphor once it has been articulated.

As George Marshall so well chronicled in his book, *Don't Even Think About It: Why We Are Wired to Ignore Climate Change*, there is a lot in the makeup and character of people as individuals and collectively as a society that makes it easy to deny the existence or implications of a changing global climate.

In this observer's view, we should not be fooled into thinking that suddenly North America is going to act on the climate threat simply because of what was agreed to in Paris. Changing people's attitudes is quite a different matter from changing their behaviour. We as a society are deeply entrenched in habits and expectations that have been centuries in the making. It is human nature not to want to lose that which, once possessed, now possesses us. Our way of life is not something we will readily give up.

While we may well be prepared to admit that the climate is changing, the conversation over what we have to do, change or sacrifice is not going to be easy. One of the reasons is that people do not like talking about issues over which they may fundamentally disagree with neighbours and even family. As Marshall points out, we avoid divisive topics we don't need to bring up, so we can get along together as colleagues, friends and relatives. He explains that some people don't want to talk about climate disruption because they just don't see it where they live. In many cases the people who feel this way live mostly indoors or inside their cars – but not everyone. I myself talk to farmers and ranchers who also just don't see it where they live. I get this. If you are used to huge natural variability in weather – temperature swings from −40° to +40°C, say, as you regularly get on the Canadian prairies – you may not see the signals.

It is one thing to intellectualize water-related climate effects but quite another to experience them directly. That breakthrough arrived for me one January night in 2003 when I finally recognized what was hidden in plain sight. In a moment of almost lightning-like illumination, I saw with my own eyes that temperatures were rising right where I live, but I had never realized it until then because these changes took place when I was paying the least attention: at night and in winter.

Others, however, don't want to talk about climate disruption because they simply don't believe it is happening and never will. The fundamental laws of atmospheric physics and our already demonstrated impacts on the global ocean notwithstanding, some people simply do not accept that humans could have that kind of impact on an entire planet.

Another source of denial relates to the nature of science itself and in particular the scientific method, which the public frequently appears not to understand. Highly publicized disputes within a divided scientific community undermine confidence in research outcomes. If you don't understand the scientific method it is easy to be distrustful of science, especially when it publicly expresses certainty about projections that don't materialize, such as the highly publicized 1990s predictions that we were about to enter a new ice age. It is clear that people like me have to do a much better job of explaining to the public that it is the very nature of science to continually pose and then resolve such questions.

Still others don't want to accept climate disruption because they don't like or trust the people who are sounding the alarm. They simply don't share the same views or values and they definitely don't trust environmentalists. This is a bigger problem than we think.

THE POLITICS OF DENIAL

While climate disruption may fundamentally be an environmental concern, it must also be taken seriously by ministries of finance, energy, economic development, infrastructure, health and, of course, agriculture. But this broadly multi-sectoral dimension of the climate threat poses its own problems. As George Marshall, for example, points out, some people deny climate change because they ideologically oppose the extent of the presence and influence government would have to exert in order to address a problem so large that it can clearly only be resolved through common, co-operative action and sacrifice. In such circumstances, what you arrive at, Marshall notes, is a form of strictly enforced silence that favours misinformation.

Ugly elements, I am afraid to say, are emerging from the tensions we have suppressed for so long over the climate threat. Research demonstrates also that there are many who don't want to talk about climate disruption because they are consciously or unconsciously troubled by the moral, ethical and legal implications of not having acted upon what they know. The central factor that determines moral and thus legal responsibility – as was demonstrated in the Memorial Hospital legal changes in the wake of Hurricane Katrina – is intention to harm. At the outset of the climate debate 40 years ago it was unthinkable that anyone would harm anyone else by consciously altering the global climate. But now, after a long series of IPCC assessment reviews and clear evidence of climate disruption on the television news every night, it is getting hard for certain interests to claim innocence when it has become clear they are, in fact, contributing significantly to the problem.

The question then becomes this: Once evidence clearly demonstrates that you are harming others by contributing to climate disruption and you still don't do anything about it, at what point does the harm you are causing become intentional? At what point does willful blindness become negligence? When does negligence become a crime?

What is interesting is that investigations such as the one the attorney general of New York initiated into the climate-change-denying activities of ExxonMobil are not related to the impacts of climate disruption per se. They revolve instead around whether or not the corporation's executive withheld from shareholders important information they possessed regarding the potential performance of the company and the true value of its assets in a world in which the continued burning of fossil fuels could potentially make the planet uninhabitable. Other oil producers, however, have complied with shareholder demands for clarification of the true value of company assets in an economy where alternative energy is clearly the future.

It is easy to see why many politicians have preferred to remain silent on the climate question. Legal issues with respect to climate impacts, however, are not going to go away. The public trust doctrine, which holds that certain resources are owned by all and available to all, is gaining legal force. Then there is the growing legal movement toward a constitutional right to a livable environment. This will demand that governments fulfill their "restorative duty," which will mean they will not just be charged with preventing future damage but will have a fiduciary responsibility to repair past harms that scientists now identify as posing a threat to current and future generations. Community leaders may want to think about that. In the middle of all this is the average person who would just like to have the opportunity to carry on living. They remain silent, hoping it will all go away. But it won't.

It is becoming clear that the climate issue could tear our society apart. What we have to avoid is falling pell-mell – everyone for themself – into a future we neither intended nor desire. It is the duty of everyone concerned about this issue to prevent this from happening. We are not going to be able to keep secret much longer the undeclared civil war that politicians are trying to contain. The silence can't last. Why? Hydro-meteorological change.

As Gudmund Hernes notes, changes in the way people think about the world do not come about as a result of a slow, cumulative process of careful observation, intellectual reasoning, sharing of knowledge and gradual building of thoughtful consensus. Society advances instead as a result of deep shocks that shatter illusions and inspire flashes of new insight that create turning points for common perspective, after which the world is never the same. Between these "moments of truth," as Hernes calls them, public attention to important issues often wanes, with momentum in the direction of change waning with it and sometimes even being lost altogether. It is only over time that the lessons learned from an accumulation of world-changing events gradually solidify into new narratives about ourselves and our future. As both George Marshall and Gudmund Hernes have clearly shown, the back and forth that takes place before a transformation occurs in public sensibility can be frustrating to contemplate and painful to experience. In the case of the climate threat, it is not just human nature that may have to change but also many of the habits we fall into in our relations with one another. Long-established, well-understood social processes stand in the way of change.

"REGULATORY CAPTURE" AND "CULTURES OF COMPLICITY"

Hernes notes there is an important class of such processes that deserves a great deal more attention. In this class he includes associations known as "unholy alliances": formal relations between regulators and the regulated that fall in the category of "regulatory capture," and collusion among parts of society that can be characterized as "cultures of complicity." Hernes defines unholy alliances as secret coalitions of outwardly opposed parties that covertly pull together when common interests are threatened or when an opportunity presents itself for hidden gain which only they have identified. In this context the military-industrial complex and the cabal of big oil, big agriculture, global food giants and automobile manufacturers come to mind, as does the elite club of transnational corporations and in particular global financial institutions.

Hernes traces the term "regulatory capture" back to 1913 and US President Woodrow Wilson, who warned that increased government regulation aimed at controlling American business would result in big business cozying up to government with the goal of capturing it in order not to be constrained by it. Evidence of this is now widespread but perhaps was most recently obvious during the administration of George W. Bush, when in the wake of the Deepwater Horizon disaster in the Gulf of Mexico it was claimed that the government regulator responsible for environmental safety in such circumstances and the regulated – in this case British

Petroleum – virtually "finished one another's sentences" when they met over matters of drilling safety practices. This happens far more often than the public knows. It is to this end that the practice of lobbying – also known as influence peddling – dedicates itself, with the aim of obtaining preferential treatment for corporations and other economic interests. In North America influence peddling is not only legal but a central feature in democratic decision making and one that politicians ignore at great risk to their electability. The problem, of course, is that once regulatory capture takes place, whether associated with deep-water drilling in the Gulf of Mexico, the allocation of water for irrigation interests, the granting of clear-cut logging licences or the environmental monitoring of oil sands operations, it is very difficult to reverse. What suffers ultimately is the public interest. What is also diminished is the capacity to respond in any meaningful way to change – especially climate change.

Even more dangerous than regulatory capture are the invisible cultures of complicity that are found everywhere at the nexus of money and power. Cultures of complicity may begin with regulatory capture but become deeper alliances that wed favoured corporations, civil servants and politicians to collusion, faulty oversight and tacit and explicit encouragement and assent to political decisions and actions that clearly violate the public trust and, in some cases, actually threaten the well-being or safety of others. The final end of a culture of complicity is outright corruption.

Cultures of complicity are seldom recognized for what they are until large-scale accidents such as chemical-plant explosions, dam failures, pipeline leaks, fishery collapses or increases in the number of climate-related disasters expose the silhouette of collusion. Until that silhouette is exposed, however, cultures of complicity invisibly harden the almost impermeably thick skin of the status quo.

Hernes goes on to outline how to identify the specific mechanisms through which an industry or an economic sector may be able to capture regulators in ways that lead to unholy alliances and cultures of complicity. These processes often begin simply with something as innocent as a common outlook across an industry as to the goals of a given sector and the benefits that sector may contribute to the common good. The regulators and the regulated are often like-minded and share a common body of knowledge. Even though the regulator is supposed to play the role of watchdog, it becomes more and more difficult, the more the knowledge and shared experience of watchers and watched converge, for outsiders to catch up with and understand the discourse between the two or to penetrate the esoteric depths of those shared values and experiences. The mere acceptance and validation of shared knowledge eventually condemns both the regulator and the regulated to relying on the same set of basic assumptions and doctrines. This is what commonly happens when, for example, the majority of those involved in the regulatory process are engineers or subscribe over time to fundamental engineering principles or perspectives. It does not take long for outsiders who do not possess the same skill set or knowledge organization principles to be consciously or unconsciously judged as not having the qualifications or perspective that would enable their views to be accepted as relevant or valid. Hernes notes – quite accurately in this observer's experience – that in such instances critics of the status quo can be privately branded extremists or stigmatized as crackpots subject to social and professional isolation and even reprisals within the camps of both regulator and regulated.

Another force that can push regulatory processes down the slippery slope of industry capture relates to who benefits from the financial success of any given economic sector. Those who gain from growing sales or increasing profits are not likely to willingly accept stricter environmental regulation, let alone countenance pronouncements that sunset is inevitable for entire economic sectors that are now currently highly profitable and employ huge numbers of people, such as the fossil fuel industry. Well-funded, highly effective lobbies exist to prevent and slow regulatory momentum in any direction that might counter the status quo. Such lobbies are far more effective than poorly organized public protests that aim to call attention to the crumbling edifice of contemporary economic and political structures in the face of rapid hydro-meteorological change.

An additional sure sign of regulatory capture, and perhaps of cultures of complicity, are the criss-crossing career paths that create overlapping opportunities for connected people in the know to work eventually for both the regulators and the regulated. Evidence suggests that this results in laxer regulations and poorer enforcement as individual regulators become less and less likely to take strong measures to control the activities of potential future employers, while former industry experts who are now in government side with their former employers on critical matters of regulation.

There is yet another very serious issue related to regulatory capture and the inadvertent creation of cultures of complicity: cuts made to the capacity and veracity of government agencies. When regulatory bodies do not have adequate in-house staff, technical expertise or experience, they become dependent over time on the very industries they are mandated to regulate, to create the metrics and mechanisms of assessment that give regulations meaning and effectiveness. In such circumstances industries literally become self-regulating and compliance is reduced to self-reporting to skeleton government agencies. This often results in weak oversight, unenforced regulations, exemptions and extensions to avert costly actions or avoid triggering public panic and blatant cover-ups.

As Hernes observes, a telltale symptom of this kind of deep complicity is the obstruction of independent media access to critical data or perspectives, which leads to under-reporting of risks. Add to all this a culture of fear created by the isolation and exclusion of critics, the ostracizing of whistle-blowers and clear threats to career advancement typical of a "you are either with us or against us" management regime, and even good, competent people toe the line. Combine this next with the underfunding of independent academic research, or restricting funding to only that research which serves the needs of industry, and it becomes possible to hermetically seal the status quo behind a perfect wall of complicity. The only way this seal can be broken is through the action of expert international or transnational agencies constructed around independent, evidence-based analysis. In terms of climate disruption, the United Nations Intergovernmental Panel on Climate Change is an example of such an agency.

Hernes would have society be more aware of the powerful role that identifying inconsistencies in the way we think and act plays in changing attitudes and behaviour. Harmonizing thinking and action, according to Hernes, can be accomplished by avoiding the trap of thinking that beliefs and values are by their nature hierarchically ordered and organized and therefore ordained. Harmonization also occurs when choices of action are identified as inconsistent or at variance with professed personal or organizational principles. Finally, it is important to recognize that attitudes align horizontally and not just hierarchically. Hernes offers this example: If you believe the global climate is being disrupted as a result of warming, and that this is happening as a consequence of human activity and is going to have a detrimental effect on the human condition in the future, then wouldn't it be reasonable for you to support measures that would attempt to do something about it?

Apart from the fact that changing your opinions may entail finding new friends, when it comes down to it, deciding to take climate change seriously is little more than an act of making what is happening in the world add up. Even if we just do the math in our heads, it is clear that despite what others around us may think, what scientists are telling us, what we are seeing with our own eyes and what we are doing about what we see don't add up at all. All it adds up to is the potential for societal collapse, collapse that at all costs we must and can avoid.

CHAPTER 7

Triggers of Transformation:
The Need for a New Narrative

We know the direction we have to go, but there is often a huge gulf between what we should do and what we actually deliver. To create a sustainable society, we need to triumph over human weakness with respect to decision making. We are not the first society to suffer from this form of internal contradiction. The Greeks actually coined a word that describes the gulf that often exists between what we know needs to be done and what we actually do – the difference between the decision we need to make, if you will, and the decision we actually render. The word is *akrasia*. *Akrasia*, according to the Greeks, was a weakness of will that resulted in "knowing what is right and failing to do it." As this book has repeatedly noted, nine Earth system boundaries have been identified as critical in that the extent to which they are not crossed marks the safe zone for human presence on this planet. Of these nine boundaries, we have already crossed the safe thresholds for biodiversity loss and the cycling of nitrogen and phosphorus through the Earth system. We also appear to be close to crossing with the changes we have made to the surface of the planet by way of altered land use and ground cover as they relate to climate change. It is important to note, however, that we know so little about the other five Earth system boundaries that it is not out of the question that we could pass over invisible and irreversible thresholds in these domains without knowing we have done so until we feel the effects. This fact in itself makes us extremely vulnerable. We are in trouble. While the crossing of these thresholds poses serious threats to the stability of the Earth system, the biggest threat of all may not be environmental deterioration, as such, but our collective response to it.

Our world at present is an event-driven one in which public perception is changed seldom by the force of argument but often by the force of events. We only alter our collective mindset when we are confronted with circumstances that overwhelm us; when what is happening around us is beyond what we have ever experienced before; when what happens to us or to people we know suddenly becomes too poignant to be denied and too disturbing to be ignored. The disruptions to date, Hernes notes, have been bad, but not bad enough to transform our view of the seriousness of the threat we now pose to ourselves. Events to date have not shaken us to the point of societal change, but because they will be beyond anything humans have ever experienced, it is only a matter of time before they do. And when they do our changing circumstances will be of such a magnitude that they will demand that fundamental societal beliefs change and that our social, economic, legal and political relations be reconstructed.

The loss of relative global hydro-meteorological stability and the new world view it demands if we are to survive the impact of this on our human condition will require as fundamental a shift in perspective as occurred when we discovered that the Earth revolves around the sun and not the converse. The object should be to get as many of us as possible through this transition as quickly as we can. The problem is that if you live in an event-driven milieu, the events that could take you out economically or environmentally over time may have already happened before you figure out how important they are and can organize to act on their larger implications. Broader narratives must therefore be written. As author William Kittredge once said about the American West:

> What we need most urgently is a fresh dream of who we are – one which tells us how to act; new stories about taking care of what we have that drive us to take appropriate action. That story, when found, will be a gift, passing from one person to another. And then our institutions will change almost at once.

Those words are now true for the entire world. Each generation needs to recreate its perspective. This is not just an iterative exercise the young need to undertake in order to find their place in the world. Rather, it is a critical process of reinvention and adaptation that is essential to making a meaningful future possible. It appears we have grown and thrived as a society in the most stable climate interval in the past 650,000 years, but that era is over. Our future will depend on our ability to confront our fate, not with panic, anger or denial but with patience, thoughtfulness and wisdom. Whatever will come to pass will not be the result of intentions, however, but of actions. As Roy Scranton has pointed out, if we want to survive in the Anthropocene, we can't just see the light; we have to *be* the light. To be the light, we need a new narrative.

THE "NATURALIST" NARRATIVE

In the book he co-edited with Clive Hamilton and François Gemenne, *The Anthropocene and the Global Environmental Crisis*, Christoph Bonneuil explores the competing narratives of our time. He begins by outlining what he calls the "naturalist narrative," in which humanity is seen to evolve from hunter–gatherers to become a global geological force. This narrative, Bonneuil observes, is marked by the view of humanity as an undifferentiated force altering the entire Earth system. It is also marked by the emergence of a global environmental conscience brought about by the capacity to monitor Earth system function, which has resulted in humanity waking up from a centuries-long dark age of unconscious impacts on the planet's biodiversity-based life support system. As a result of this awakening there has been a consequent rise in the role of scientists as "shepherds" of humanity and of trust in technology as a means of escaping responsibility for the damage we are causing.

Given that in only a few centuries humanity has become enough of a geological force to have our own impacts represented as an anomaly in stratigraphic nomenclature, some commentators are inclined to call this narrative the Manthropocene. From this we can see that it is not all about climate. It is that we have changed the composition of the Earth's atmosphere by adding not just carbon dioxide but also other substances. We have altered the chemistry of our oceans, interrupted the water cycle, altered the character of more than half of the land surface of the Earth and are on our way to wiping out half of the other creatures big enough to see with which we share our planet. On top of this, we have changed the climate.

Bonneuil notes there are many problems with the naturalist narrative, not the least of which is that, while they may not have the capacity to express it, the rest of creation – including all the species whose very existence is now threatened – will not have the same view of the Anthropocene. Another problem is that we humans are not an undifferentiated biological agent. Some of us cause a lot more damage than others. The Anthropocene, Bonneuil points out, is a product of a combination of highly directed belief systems, socio-technological imperatives and economic and political choices. We didn't just blunder innocently into this new era. The Anthropocene is a thoughtless crime we consciously committed. Despite what the naturalist narrative might imply, we didn't stumble into our current circumstance for want of warnings concerning the potential consequences of our impacts. Thoughtful observers have been issuing caveats about our potential impacts for centuries. It is chilling now to realize that parliaments in Western Europe were regularly debating the ultimate effects of deforestation and environmental degradation, as well as consequent climate effects, as early as 1780. As Bonneuil notes, the standard narrative – that until the past 50 years or so, only local knowledge about environmental impacts applied in the absence of systematic knowledge of global environmental change – simply doesn't hold water. We entered the Anthropocene in spite of that knowledge.

THE "POST-NATURE/ END OF NATURE" NARRATIVE

Bonneuil also offers another narrative, which in the present author's opinion is the most frightening of all. In this world view, science is lifting the veil of past environmental ignorance that characterizes the grand narrative of the naturalists. According to this notion, science, and in particular engineering, will now guide humankind and be the salvation of the world. It is an approach that presses for internationally accepted, large-scale geo-engineering to optimize the global climate. This "post-nature/end of nature" narrative, Bonneuil argues, pictures society as ignorant, passive and mired in cognitive dissonance. The solutions are clear: if we have arrived at the end of self-willed, self-regulated nature, scientists must take the lead and imagine new Earth-system-replacing technologies. In this scenario, techno-utopians and geo-engineers promise us a world with no need for nature in an ever more prosperous Anthropocene. They say we need to enthusiastically embrace technologies, including those that were once taboo such as nuclear power, genetic engineering and geo-engineering, to assure humankind's future. Technological risks are normalized and become part of the human condition. As Bonneuil puts it, to achieve a "good result" in the Anthropocene, all we have to do is embrace, love and improve on our technological Frankensteins.

In this narrative, the concept of "nature," as opposed to the understanding of the physical and chemical workings of natural systems, is seen as merely a human construction shaped by human desires and ends. A new eco-pragmatism is now required that rejects Romantic notions of what is natural. Critics of this narrative argue that geo-engineering is part of the exhausted "man conquering nature" scenario that we have to transcend. Giving Frankenstein's monster regular facelifts will not get us there. Critics also point out that pronouncements of the death of nature may be premature.

The notion that nature as we have known it no longer exists is contested by those who have observed the capacity of natural systems to regenerate, and by those who reject technological fixes as a means of saving the planet. Critics of this narrative also argue that all the post-nature narrative does is intensify and accelerate the current, exhausted idea of modernity. We need a different grand narrative, one that embraces the rethinking of what individual freedom means and that gives political existence to the rest of creation.

THE "ECO-CATASTROPHE" NARRATIVE

A third grand narrative is the one put forward by those Bonneuil identifies as "eco-catastrophists." In this story, unsustainable agricultural practices, resource depletion and transgressed Earth system boundaries lead to tipping points and step-like changes in vital Earth system function. In this approach, the Anthropocene is an epoch in which infinite growth and relentless progress hit the wall of the planet's finitude.

The eco-catastrophist narrative acknowledges the possibility of a collapse of the industrial way of life, and the acceptance of limits to growth becomes an opportunity for more participatory politics and a new, simpler, more local, post-growth society. The eco-catastrophist brief argues for the urgent need to radically change the dominant way of living based on producing and consuming, and rejects the belief that technological fixes will save the planet within the frame of unchanged social, political and economic circumstances. In this scenario, science and technology alone cannot "save the planet"; social change is necessary if we desire to create a sustainable and adaptive society.

THE "ECO-MARXIST" NARRATIVE

This approach sees the Anthropocene in the context of the Marxian view that an ultimately fatal contradiction in capitalism resides in its inability to maintain natural systems. In this idea, Bonneuil notes, the Anthropocene is a symbol of the unsustainable metabolism of the capitalist system, brought about by internationalization of the economy and gross asymmetries in the division of labour. Rather than people bringing about the Anthropocene, it is capital that did the dirty deed. The Anthropocene is viewed as the "Capitalocene." What is missing from this notion, however, is the fact that it is carbon, not capital, that has to be the new currency in the Anthropocene.

CARVING A WAY FORWARD

As it now appears, we are heading for a huge fight over what the guiding narrative for humanity's future will be. This ground is already being staked. On one hand we have the techno-utopians and the geo-engineers who want us to love our Frankensteins. On the other are those who believe that pronouncements of the death of nature may be premature. On one side are those who believe that normalizing technological risks is now a necessary part of the human condition. On the other are voices like Kathleen Dean Moore, who want us to begin to rethink the very principles of what it means to be human by first granting natural objects, other species, ecosystems and natural cycles the rights of personhood, including the rights of protection and restoration. Moore urges society to move in the direction of adaptation strategies that do not simply stiffen our present habits and attitudes with steel beams by merely reinforcing current patterns of excess and exploitation. According to Moore, the choice we have to make now is between self-delusion and self-fulfillment through creating not a new world but a better one.

In the middle of all this debate is the average person who simply wants to get on with their life. In the face of so much terrifying news they feel helpless. "I am just one person, what can I do?" they ask. Moore has an answer to that question. One thing you as one person can do, she offers, is stop being just one person. Look around you. You are not alone. You may not be able to change *the* world, but you can change *your* world and in so doing influence others around you.

We should not wait for a huge shock to a large number of people to move public attitudes in the direction of a new narrative. We have transformed the world, and now the world is about to transform us. What we are engaged in is nothing less than a struggle to redefine our dominant mythology. Given our circumstances, we might agree with William Kittredge that finding this story should be our society's most urgent common enterprise.

Moore offers that it is time to speak out. It is time for relentless, informed, fearless and courageous citizenship. She offers that water can be our guide. One of the ways we can get through this difficult bottleneck in the existence of humanity is to think like a river. In rivers no single flow pattern persists indefinitely. Nor does any river ever stay the same. When, after much deposition of sediments flowing from upstream, the course of the river is so obstructed by sandbars and islands that a channel can no longer carry all its water and sediment, rivers can cross a threshold of stability that causes its current to carve in a new direction. In a process hydrologists call avulsion, rivers can change their course suddenly and dramatically. This is what we need society to do now. If we can establish a common vision and work together, our small individual influences on the flow of the river of time can cause huge and sudden changes. Knowing that we rely on biodiversity for our very survival, whatever we decide to call the new epoch we have entered, the goal should be to create a more equitable, just and sustainable future for all life on this planet.

BOOKSHELF

Bonneuil, Christophe. "The Geological Turn: Narratives of the Anthropocene." In Clive Hamilton, Christophe Bonneuil and François Gemenne, eds. *The Anthropocene and the Global Environmental Crisis*, c2.

Crutzen, Paul J., and Eugene F. Stoermer. "The 'Anthropocene.'" *Global Change Newsletter* 41 (2000): 17–18. Accessed 2016-04-05 (pdf) at https://is.gd/U6Qz53.

Fink, Sheri. *Five Days at Memorial: Life and Death in a Storm-Ravaged Hospital*. New York: Crown Publishers, 2013.

Hamilton, Clive, Christophe Bonneuil and François Gemenne, eds. *The Anthropocene and the Global Environmental Crisis: Rethinking Modernity in a New Epoch*. London and New York: Routledge, 2015.

Hernes, Gudmund. *Hot Topic – Cold Comfort: Climate Change and Attitude Change*. Oslo: NordForsk, 2012.

Hsiang, Solomon, and Amir Jina. "The Causal Effect of Environmental Catastrophe on Long-Run Economic Growth: Evidence from 6,700 Cyclones." National Bureau of Economic Research Working Paper No. 20352, July 2014. Accessed 2016-04-05 (pdf) at www.princeton.edu/rpds/events_archive/repository/Hsiang030415/Hsiang030415.pdf.

Hutton, Guy. "Can We Really Put a Price on Meeting the Global Targets on Drinking-Water and Sanitation?" *The Water Blog*, February 12, 2016. Washington, DC: World Bank. Accessed 2016-05-01 at http://is.gd/8nFkwH.

Hutton, Guy, and Mili Varughese. "The Costs of Meeting the 2030 Sustainable Development Goal Targets on Drinking Water, Sanitation and Hygiene." World Bank Group Water & Sanitation Program Summary Report 103172. Washington, DC: World Bank, January 2016. Accessed 2016-04-05 (pdf) at http://is.gd/vxywR2.

Jamieson, Dale. *Reason in a Dark Time: Why the Struggle against Climate Change Failed and What It Means to Our Future.* New York: Oxford University Press, 2014.

Kittredge, William. *Taking Care: Thoughts on Storytelling and Belief.* Minneapolis, Minn.: Milkweed Editions, 1999.

Koestler, Arthur, and Herbert Butterfield. *The Sleepwalkers: A History of Man's Changing Vision of the Universe.* London: Hutchinson, 1958.

Marris, Emma. *Rambunctious Garden: Saving Nature in a Post-Wild World.* New York: Bloomsbury, 2011.

Marshall, George. *Don't Even Think About It: Why We Are Wired to Ignore Climate Change.* New York: Bloomsbury, 2014.

Moore, Kathleen Dean. *Great Tide Rising: Towards Clarity and Moral Courage in a Time of Planetary Change.* Berkeley, Calif.: Counterpoint, 2016.

Schuessler, Ryan. "This Could Explain All Those Strange Happenings in Alaska's Waters." *Washington Post*, February 16, 2016. Accessed 2016-04-05 at www.washingtonpost.com/news/energy-environment/wp/2016/02/16/

this-could-explain-all-those-strange-happenings-in-alaskas-waters.

Schuster-Wallace, C.J., and R.W. Sandford. *Water in the World We Want*. Hamilton, Ont.: United Nations University Institute for Water, Environment and Health and United Nations Office for Sustainable Development, 2015. Accessed 2016-04-05 (pdf) at http://inweh.unu.edu/wp-content/uploads/2015/02/Water-in-the-World-We-Want.pdf.

Scranton, Roy. *Learning to Die in the Anthropocene: Reflections on the End of a Civilization*. San Francisco: City Lights Books, 2015.

Slade, Giles. *American Exodus: Climate Change and the Coming Flight for Survival*. Gabriola Island, BC: New Society Publishers, 2013.

Steffen, Will, et al. *Global Change and the Earth System: A Planet Under Pressure*. Berlin: Springer, 2004.

United Nations. *The Paris Agreement under the Framework Convention on Climate Change*. Adopted by the 21st Session of the Conference of the Parties, Paris, December 11, 2015. Accessed 2016-04-05 at https://sustainabledevelopment.un.org/frameworks/parisagreement.

United Nations. *Transforming Our World: The 2030 Agenda for Sustainable Development*. New York: UN General Assembly Resolution A/RES/70/1, September 27, 2015. Accessed 2016-04-05 at https://sustainabledevelopment.un.org/post2015/transformingourworld.

United Nations Conference on Trade and Development. *Wake Up Before It Is Too Late: Make Agriculture Truly Sustainable Now for Food Security in a Changing Climate*. UNCTAD Trade and

Environment Review 2013. New York: United Nations, 2013. Accessed 2016-04-05 (pdf) at http://is.gd/hyonI8.

Wilson, E.O. *Half-Earth: Our Planet's Fight for Life*. New York: Liveright Publishing (W.W. Norton), 2016.

World Bank. "The Costs of Meeting the 2030 Sustainable Development Goal Targets on Drinking Water, Sanitation and Hygiene." See Hutton, Guy, and Mili Varughese above.

ABOUT THE AUTHOR

Robert William Sandford is the EPCOR Chair for Water and Climate Security at the United Nations University Institute for Water, Environment and Health. In addition, Bob is the author of 30 books on the history, heritage and landscape of the Canadian Rockies, including *Water, Weather and the Mountain West* (RMB, 2007), *The Weekender Effect: Hyperdevelopment in Mountain Towns* (RMB, 2008), *Restoring the Flow: Confronting the World's Water Woes* (RMB, 2009), *Ethical Water: Learning to Value What Matters Most* (RMB, 2011), *Cold Matters: The State and Fate of Canada's Fresh Water* (RMB, 2012), *Saving Lake Winnipeg* (RMB, 2013), *Flood Forecast: Climate Risk and Resiliency in Canada* (RMB, 2014), *Storm Warning: Water and Climate Security in a Changing World* (RMB, 2015), *The Columbia Icefield – 3rd Edition* (RMB, 2016) and *North America in the Anthropocene* (RMB, 2016). He is also a co-author of *The Columbia River Treaty: A*

Primer (RMB, 2015) and *The Climate Nexus: Water, Food, Energy and Biodiversity in a Changing World* (RMB, 2015). Robert lives in Canmore, Alberta.

The RMB *manifestos*

PASSIONATE. PROVOCATIVE. POPULIST.

RMB has created one of the most unique non-fiction series in Canadian publishing. The books in this collection are meant to be literary, critical and cultural studies that are provocative, passionate and populist in nature. The goal is to encourage debate and help facilitate positive change whenever and wherever possible. Books in this uniquely packaged hardcover series are limited to a length of 20,000–25,000 words. They're enlightening to read and attractive to hold.

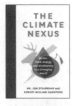

THE CLIMATE NEXUS

Water, Food, Energy and Biodiversity
in a Changing World

Jon O'Riordan & Robert William Sandford

ISBN 9781771601429

THE EARTH MANIFESTO

Saving Nature with Engaged Ecology

David Tracey

ISBN 9781927330890

ETHICAL WATER

Learning To Value What Matters Most

Robert William Sandford & Merrell-Ann S. Phare

ISBN 9781926855707

FLOOD FORECAST

Climate Risk and Resiliency in Canada

Kerry Freek & Robert William Sandford

ISBN 9781771600040

THE COLUMBIA RIVER TREATY

A Primer

Robert William Sandford, Deborah Harford & Jon O'Riordan

ISBN 9781771600422

SAVING LAKE WINNIPEG

Robert William Sandford

ISBN 9781927330869

RMB saved the following resources by printing the pages of this book on chlorine-free paper made with 100% post-consumer waste:

Trees · 8, fully grown
Water · 3,680 gallons
Energy · 4 million BTUs
Solid Waste · 246 pounds
Greenhouse Gases · 678 pounds